# APPLIED FISHERIES

## ABOUT THE AUTHOR

**Dr. Vishwas Balasaheb Sakhare**, born in the year 1974, took his M.Sc.and Ph.D. degrees from Swami Ramanand Teerth Marathwada University, Nanded (Maharashtra) and postgraduate course Inland Fisheries development and administration from Central Institute of Fisheries Education, Kolkata.Dr.Babasaheb Ambedkar Marathwada University; Aurangabad (Maharashtra) selected Dr.Sakhare as a research guide in the subject of Zoology.

Dr. Sakhare is presently working as lecturer in Post Graduate Department of Zoology, Yogeshwari College, Ambajogai (Maharashtra). Besides teaching, he has done significant work on reservoir fisheries management and limnology and has published 22 research papers in national and international journals, apart from this, over 30 popular articles in science magazines and newspapers. He is the recipient of fellowship of the Indian Association of Aquatic Biologists.

Dr. Sakhare is also an active member of several scientific societies. He is working as an associate editor of *'SAROVAR'* a newsletter of South East Asia Network on Lakes and Reservoirs (SASANET L and R). He is a well-known fisheries consultant providing technical guidance to fisheries co-operatives engaged in reservoir fishing. He has participated in many symposia and workshops.Dr.Sakhare is also a recognized post-gradate teacher of Dr.Babasaheb Ambedkar Marathwada University. He is the author of some books like *'Ecology of Lakes and Reservoirs'* and *'Reservoir Fisheries and Limnology'*.

# APPLIED FISHERIES

**Dr. Vishwas B. Sakhare**
*Post Graduate Department of Zoology*
*Yogeshwari Mahavidyalaya,*
*Ambajogai – 431517*
*Maharashtra*

**2021**
**Daya Publishing House®**
*A Division of*
**Astral International Pvt. Ltd.**
**New Delhi - 110 002**

ISBN: 9789388173896 (Hardbound)

*Reprinted*, 2021



*Published by* : **Daya Publishing House**
*A Division of*
**Astral International Pvt. Ltd.**
– ISO 9001:2015 Certified Company –
4736/23, Ansari Road, Darya Ganj
New Delhi-110 002
Ph. 011-43549197, 23278134
E-mail: info@astralint.com
Website: www.astralint.com

***Printed at*** : **Replika Press Pvt. Ltd.**

# Preface

This book is based on recent views, ideas, and progress in the field of fisheries. An attempt is made to cover some important aspects of fisheries. It covers topics such as live feed organisms of prawns, culture and breeding of magur and murrels, impact of exotic fishes on indigenous fish fauna, fisheries cooperatives, marketing etc. I have tried to present these aspects in an easy manner to read for teachers, students and researchers.

I am grateful to all the scientists and workers who supplied the requisite literature and information.

Iam highly thankful to Dr. Sureshji Khursale, President, Yogeshwari Education Society, Dr. D.H. Thorat, Principal, Yogeshwari Mahavidyalaya, Ambajogai and all the staff members of Yogeshwari Mahavidyalaya, Ambajogai for their encouragement.

I am grateful to Dr. M.S. Kodarkar, Dr. P.K. Joshi and Dr. R.J. Chavan for their encouragement.

I am deeply grateful to my family members for their forbearances they endured partial loss of my company when I was busy with the manuscript. The author is much indebted to Miss Jayshri Shingare for her expertly drawn figures.

I thank Mr. Anil Mittal of Daya Publishing House, Delhi for his cooperation in production of this book.

Thanks are also due to all those who have not been mentioned by name but nevertheless have been of invaluable help in their inscrutable ways.

Inspite of care bestowed, there may be some mistakes and defects and I feel grateful to all friendly critics, who help me in improving the text.

**Dr. V.B. Sakhare**

# Contents

*Chapter 1*

# Live Feed Organisms in Prawn Hatcheries

Availability of appropriate feed is one of the major factors in the culture of larvae of prawns, as it is directly responsible for their growth and survival. The demand for seed of cultivable species has increased considerably during past two decades. Further expansion and sustenance of this industry rests on production of quality seeds. But availability of suitable feed for larval stages is still a bottleneck in commercial seed production. A variety of artificial pelleted feeds are available to rear larval stages of cultivable species, which are not preferred, compared to live food (Sumitra, 1987). Live feed is superior to compound feed because it is readily ingested (Kinne, 1977), digested more easily, does not affect water quality (Watanabe *et al.*, 1978) and has essential growth factors. Live food organisms contain digestive enzymes and there by assist in the digestive process (Dabrowskii and Glogowski, 1977). They have high food value as protein source (Watanabe *et al.*, 1983) and amino acid richness (Ravi and Devaraj, 1991) for cultivable species.

Live feed can be classified into two types *i.e.*, algal and nonalgal. Algal feeds include all phytoplankton, marine yeast and bacteria. The nonalgal feed consists of zooplankton. Penaeid prawn larvae require different types of feed during their development. These requirements vary from purely algal feed, mixed feed (algae and zooplankton) to purely zooplankton.

## Algae

Different algal species have been tried in the protozoeal stages of penaeid prawn. The species popularly used are *Chaetoceros sp, C. gracilis, C.calcitrans, Thalassiosira weissiflogii, Nitzchia closterium, Cylindrotheca sp, Tetraselmis sp, Chlorella sp, Isochrysis galbana* etc.

Algal food can be cultured along with the prawn larvae in the rearing tank (same tank method) or separately in the indoor conditions (separate tank method). In the same tank method a mixed diatom is generally cultured with the aid of sunlight and by suitable fertilization. In addition the algae stimulate the growth of rotifers and copepods, and form the natural food. In separate tank method a monoalgal culture is maintained separately and provided to the prawn larvae every day.

Even though the first successful rearing of prawn larvae (*P.japonicus*) was obtained in Japan by employing the separate tank method. The same tank method was gradually replaced the separate tank method. This method has been adopted in many countries. In this method algae are cultured in rectangular concrete tanks of 15 to 250 ton capacity. The propagation of diatoms in the tanks is controlled by adjusting the quality of the fertilizer or by varying the illumination density. The commonly used fertilizers are potassium nitrate and dibasic potassium phosphate. Several algal species contribute to the production, but among them *Skeletonema costatum, Melosira sp, Nitzchia sp, Rhizosolenia sp* were the dominant diatoms, overblooming of algal cells is the main problem facing the same tank method, sudden collapse of the algal bloom in the afternoon causing mortality of the larvae. It is assumed that the substances released by the death of the diatom cells are toxic to the larvae.

The method followed in India shows a similarity to both the same tank and separate tank method.

There is no single algal species, which can be regarded as the best in every respect for culture and use as feed for larval penaeids in the world. Some algae are easily cultured and maintained, while others are not. The suitability of the species mostly depends on certain algal characteristics, the culture conditions, and local environmental factors as well as the food requirements of the prawns.

## Zooplankton

### Artemia

It is commonly known as 'brine shrimp' or 'sea monkey'. It is a crustacean thriving in hypersaline waters in which predators and competitors are few, and algal production is high. There are 50 brine strains from various continents and countries. Some of the important species of artemia are *Artemia salina* and *Artemia franciscana*. The life span of artemia is about six months. It possesses both bisexual and parthenogenetic strains. Adults measures about 10mm in length.

The life cycle begins from a dormant cyst that contains an embryo in a suspended state of metabolism. The cysts are hardy, which may remain viable for many years if kept dry. Decapsulated cysts and freshly hatched nauplii of artemia are used as food for larval and post larval stages of shrimps and fishes. Reproductively active adult artemia form as a food to induce maturity in the broodstock of shrimps and fishes. The advantages of feeding artemia are that the developing shrimps and finfish are put into a more natural environment by having to change their food. In addition, live artemia are said to provide developing shrimp and finfish with some of the best nutrients available.

Artemia were discovered in 1831 in the Great Salt Lake in the Utah desert. Artemia is the suitable fed for zoea and early post larval stages of prawns (Sorgeloos, 1980). High price, availability, variable nutritional quality and hatching efficiencies among different strains are some of the well-known negative characteristics of this live food (Watanabe *et al.*, 1982 and Leger *et al.*, 1986). Kinne (1977) has

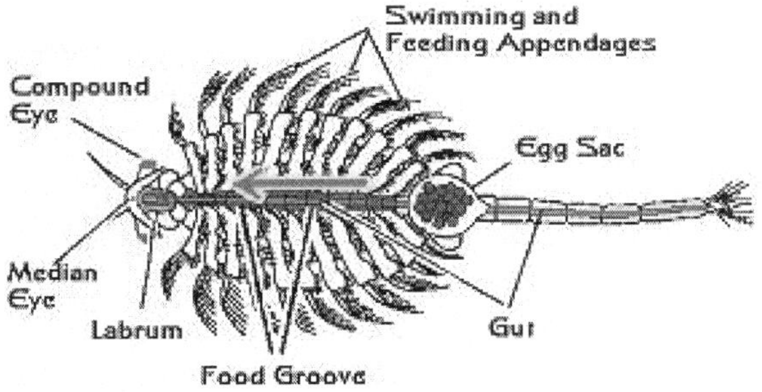


**Figure 1.1: *Artemia***

indicated that more than 85 per cent of the marine animals cultivated so far have been offered *A. salina* as feed. Hudinga in 1958 for the first time successfully reared *P.japonicus* using Artemia nauplii during mysis and post larval stages. Freshly hatched artemia nauplii appear to be a better food to the larvae of *P.monodon, P.indicus, Metapenaeus monoceros*, and *Macrobrachium rosenbergi*. The number of artemia nauplii is kept at a density of 1 to 5 nauplii per ml of culture medium. The feeding experiments conducted on *P.semisulcatus* by Al-Hajj *et al.* (1983) showed that for the development of mysis stage to first postlarval stage, an average of 183.33 artemia nauplii per larva per day was optimal. Recently the use of encapsulated cysts in the rearing the larvae of *P.monodon, M.ensis, M.endevouri* and *M.rosenbergii* and *M.monoceros* has shown good survival.

The culture of marine shrimp and finfish larvae requires premium quality artemia, high in cyst count and per cent hatchout (the number of hatched nauplii as a per centage of the total number of cysts per gram), with small nauplii and superior nutrient contents. Most importantly, the nutrient profile must provide sufficient quantities of certain highly unsaturated fatty acids (HUFAs), which inter alia are responsible for molting in shrimps. In the San Francisco Bay, the HUFA content of the plankton, which the adult artemia feed upon provides abundant sources of the critically important eicosapentaenoic acid 20: 5W3 and dodecahexonoic acid 22: 6W3.

The natural population of artemia found in India belongs to the parthenogenetic strain. They are commonly known as 'artemia parthenogenetica'. Among the cysts produced by the various strain from different parts of the world, the cysts of the Indian strain are the largest (284 mm); the smallest (225 mm) are the San Francisco Bay strain. The size of the food organism is critical during the early larval stages of shrimp, feeding a large nauplius to the small predator may result in poor growth of the predator and even mortality due to starvation. Hence, the large-sized Indian artemia strain is not suitable for feeding early post larval stages of shrimp. The Indian artemia also contains low levels of HUFA, which show up in the low survival and poor growth of shrimps.

Almost the entire artemia requirements of shrimp and freshwater prawn hatcheries in India are met from import of cysts from USA.India imports about 65-70 metric tones of artemia per annum, which includes the requirements of both shrimp and freshwater prawn hatcheries. Presently, the Government of India given permission to the importer for each consignment.

## Rotifers

Dutruchet (1812) was first to regard them as a separate biological group distinct from protozoa and called them rotifera. Over 2500 rotifer species belonging to 200 genera are known from various works in different countries. Most of the rotifers are freshwater forms. They are well known as food for fishes as well prawns. Among them *Brachinous calcyflorus* and *B.ruben* have been proposed as food for freshwater fish larvae. Only one species, *B.plicatilis* is euryhaline, having been cultured in seawater.

The rotifers are cultured on a mass scale in many hatcheries. A major breakthrough in rotifer culture was made by Hirata and Mori (1967) who found that they could be raised on baker's yeast. Later pure cultures of *Chlorella sp* were tried as feed.The rotifers have shown higher reproductive rate in *Chlorella sp*, medium them in yeast. Following this work several species of marine yeast were isolated and they were found to be adequate food for rotifers. Some marine bacteria are also used as food for rotifer. Yasuda and Taga (1980) have reported that the rotifers fed on two *Pseudomonas* strains, have shown good growth. In artificial feeds the rotifers showed better growth.

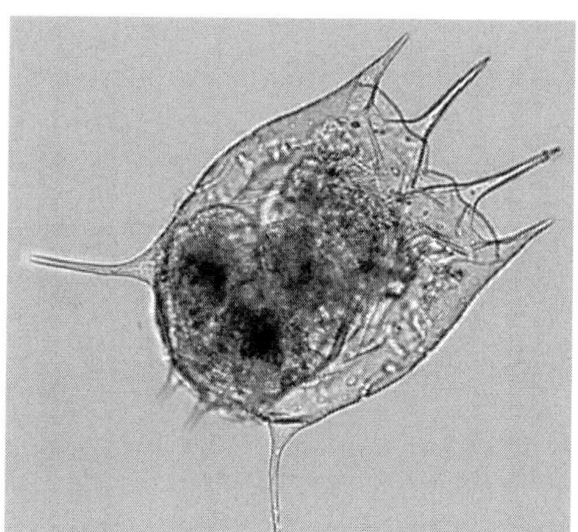

**Figure 1.2: *Brachionus calyciflorus***

*B.plicatilis* is a one of the important components of the hathchery. The densities of rotifers applied to the larval tank may range between 5 and 30 individuals per milliliter. Owing to their small size they could be introduced in the culture tanks when the larvae are in protozoeal stage. El-Amad (1982) reported that the optimum feeding level for the *P.semisulcatus* larvae may be close to a mean value of 112.3 rotifer per larvae perday. Al-Hajj *et al*. (1983) have reported for the same species that the larvae consumed on an average 240 rotifers per day between the first protozoeal and the first mysis stages. In adverse conditions, frozen rotifers were also offered to the larvae. *P. japonicus* larvae did not show any variation in growth when fed on either frozen or live rotifer.

## Copepods

Copepods are very small crustaceans. Their size varies from 0.3 to 2.5 mm and body is cylindrical or pear shaped having long appendages. Among copepods, *Cyclops, Mesocyclops, Diaptomus, Canthocamptus* etc are used by fishes as their food. Free-living copepods are an essential link in food chain occupying the intermediate trophic level between bacteria, algae and protozoa on one hand and small and large plankton predators on the other.

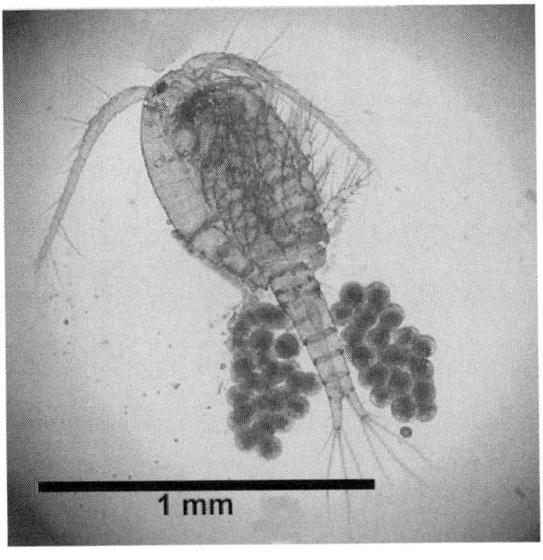


**Figure 1.3: *Mesocyclops edax***

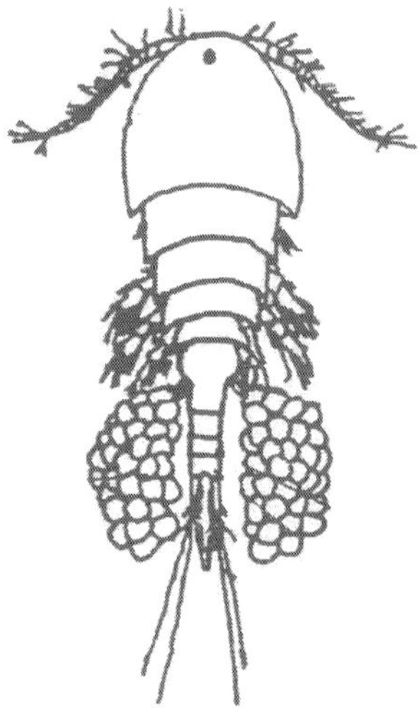

**Figure 1.4:** ***Diaptomus***

In a review on the cultivation of copepods, Omari (1973) has stated that so far only 30 species –20 calanoids, -1 cyclopoid and 9 harpacticoids have been reared for one or more filial generations under laboratory conditions.

Algae and diatoms were found to be excellent diet for the culture of copepods. Few copepods such as *Pseudocalanus minutus, P.elongates, Temora longicornis* and *Eurytemora affinis* were successfully reared on *Isochrysis galbana*. The marine yeast *Candida sp* and *Chlorella sp* were used for the mass culture of a cyclopoid, *Apocyclops borneoensis*. The experimental mass culture of *Nitocra spinipes* was carried out in four different feeds such as *Chlorella sp*, shrimp head meal, compound feeds and *Salvinia* detritus, among which *Chlorella sp* served as the best feed. Goswami (1977) based on his experiments suggested that the *Laophonate sestosa* can be reared on detritus.By using market vegetable or wheat bran, higher densities were achieved with several other species like *Tisbe holothuriae, Amphiscalla subdebilis, N.spinipes* and *Microstris elatensis*.

## Cladocerans

The greater significance of cladocera in the aquatic food chain as food for both young and adult fish was emphasized much earlier. Generally the stomach content of young fish has 1 to 95 per cent cladocera by volume. Out of 11 families of cladocera, 8 families have been reported from Indian waters, which represents about one fourth of the world cladoceran fauna. Nutrient rich cladoceran food is critical in the development of larval stages of culture fish species, as during these developmental stages gastrointestinal tract as well as protelytic enzyme systems are not well developed. Among freshwater zooplankton, the organisms belonging to the cladoceran group have been widely used in fish and shellfish hatcheries. Two important genera such as *Daphina* and *Moina* have been established as important live food organisms in culturing peanaeid larvae. Although thcy are freshwater organisms they have been successfully used in the frozen condition to feed marine organism.

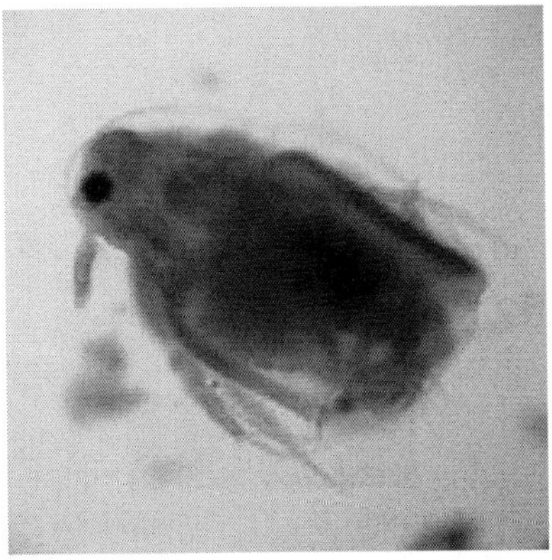

**Figure 1.5: *Moina***

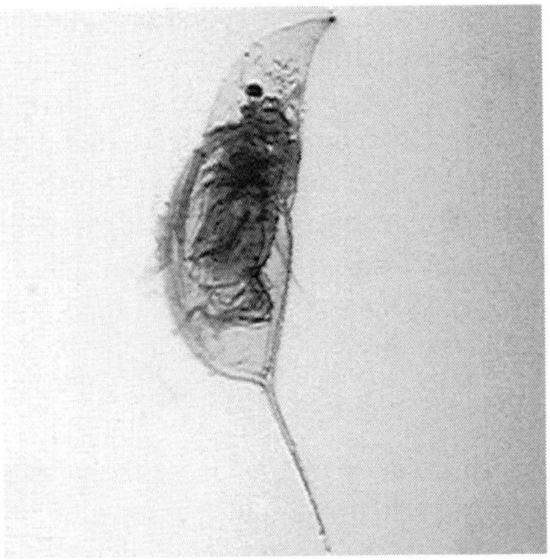

**Figure 1.6: *Daphnia retrocurva***

Banta (1921) used 'stable tea' an extract of horse manure to culture cladocerans in the laboratory. Coklin and Provasoli (1971; 1978) have successfully grown *M.macrocopa* in a synthetic biphasic medium. Experimental results of Nandy *et al.* (1977) indicate that *D.lumhotlzi* attained a maximum production of 12650 to 15000 per litre when fed with brewer's yeast, but poor production in *Chlorella vulgaris* and in a mixed diet of *Chlorella sp* and poultry manure. By using chicken manure ventura and Enderz (1980) developed *Moina sp*, culture with a density of 5000-10,000 animals per litre. De Pauw *et al.* (1980) maintained a population of *D.magna* cultured on a sole feed of micronised rice bran (less than 60 micron) for 20 continuous generations without noticeable differences.

### Nutritive Value of Food Organisms

The growth of any organism is mainly dependent on the nutritive value of the feed *i.e.*, the quantity of protein lipid and carbohydrate. An important determinant of the overall nutritional value of a foodstuff for crustaceans is its lipid content (Schaver *et al.*, 1980). The copepods *Tigriopus* and *Acagtia sp* contain relatively high amount of 20: 5W3 irrespective of the culture media and the organisms (Watanabe *et al.*, 1980). The *Daphina sp* and *Moina sp* have shown low fatty acids when fed with yeast, but gained high concentrations of 20: 5W3 by feeding poultry manure and algae, thus making them ideal for fish from the view point of EFA.

## Conclusion

With semi intensive and intensive culture of prawns now gaining ascendance in India, where the labour is cheap and the problems of pollution, which less unlike in the developed countries, economically viable live food should be encouraged by isolating and culturing locally available species. In developing countries like India, where the ambient temperature supports luxuriant growth of live feed throughout the year, the culture of these organisms may be economically and biologically viable when compared to many artificial feeds.

## REFERENCES

Al-Hajj, A.B., C.M.James, S.Al-Ablani and A.S.D. Farmur. 1983. Advances in hatchery production of *Penaeus semisulcatus* in

Kuwait with particular reference to the refinement of feeding practices. *J.World Maricul.Soc.* 14: 64–74.

Banta, A.M., 1921. A convenient culture medium for daphnids. *Science,* 53: 557–558.

Dabrowski, K. and Glogowski, 1977.The role of exogenic protelytic enzyme in digestion process in fish. *Hydrobiol.,* 57: 9–134.

Kinne, O., 1977. Cultivation of animals. In: Kinne (Ed.). Wiley, London, *Mar. Eco.,* 3 (2): 579–1293.

Leger, P., Bengtson, D.A., P. Sorgeloos, K.L.Simpson and A.D. Beck, 1986. The use and nutritional value of *Artemia* as a food source. *Oceanogr. Mar. BW Ann. Rev.,* 24: 521–623.

Neelakantan, B., A.S. Rafiuddin, N.R. Ratish Menon and N. Kusuma, 1988. Importance of live feed organisms in prawn hatcheries– A review. *J. of the Indian Fisheries Asso.,* 18: 47–67.

Ravi, J. and K.V. Devaraj, 1991. Quantitative essential amino acid requirements for growth of *Catla catla* (Hamilton). *Aquaculture,* 96: 281–291.

Sorgeloos, P., 1980.The use of the brine shrimp *Artemia* in aquaculture. pp. 25–46. In: G. Personne, P. Sorgeloos, O. Roles and E. Jaspers (Eds.). *The Brine Shrimp, Vol. 3: Ecology, Culture, Use in Aquaculture.* Universal Press, Wetteren, Belgium.

Sumitra, V., 1987. Cultivation of live food organisms status and scope in India. In: *Advances in Aquatic Biology and Fisheries,* Prof. N. Balakrishnan Nair Felicitation Volume. Natarajan, P. Suryanarayanan, H. Azis, P.K.A. (eds.), p. 239–249.

Watanabe, T., T. Arakawa, C. Kitajima, K. Fukusho and S. Fujita, 1978.Nutritional quality of living feed from the view point of essential fatty acids for fish. *Bull. Jap. Soc. Sci. Fish.,* 23: 1227.

Watanabe, T., M. Ohta, C. Kitajima and S. Fujita, 1982. Improvement of dietary value of brine shrimp *Artemia salina* for fish larvae by feeding them on omega 3 highly unsaturated fatty acids. *Bull. Jap. Soc. Fish.,* 48(12): 1775–1782.

Watanabe, T., Kitajima C. and Fujita, S., 1983. Nutritional values of live organisms used in Japan for mass propogation of fish: A review. *Aquaculture,* 3: 115–143.

*Chapter 2*

# Bheries

Bheries are of the most important fishery resources of West Bengal. This traditional fishery is reported to have developed first in the spill area of the Bidyadhari river consequent to the silting up of the river due to discharge of Kolkata city sewage. Later this fishery practice was expanded to the lower deltaic region of the Sunderbans. Loally, this fishery is known as 'Basa bada' , 'non-gheri' or 'bheries'.

Bheries have been playing an important role as fish culture media since ancient time. Fish/shrimp farmers now utilize the sewage effluents to increase the productivity of bheries (culture ponds) for culturing shrimps and fish following the traditional method. Most of the farmers follow he polyculture method with a dominance of shrimp component.

## Types of Bheries

Bheries are of two types *i.e.*, wastewater bheries and saline bheries.

Wastewater bheries are nothing but the sewage –fed bheries. These are located in places like Salt Lake, Jadavapur, Tiljola, Sonarpur and Bhangar areas of West Bengal. These are highly productive. Sewage of Kolkata city is drawn directly into bheries for aquaculture. The nutrient rich water stimulates the production of natural food items for fishes. The Indian major carps (rohu, catla and mrigal) and exotic carps (silver carp, grass carp, common carp and tilapia) are mainly cultured in bheries.

Saline bheries are mainly located at Barasat, Basirhat, Rajarhat, Deganga in low saline area, Minakhan, Haroakulti, Malancha in medium saline area and Nazat, Basanti, Gosaba, and Kakdwip in high saline region. Tidal waters are drawn into bheries from rivers and their tributaries. In these bheries fresh as well as brakishwater fish and prawn are cultured.

## Management of Bheries

During the offseason bheries are properly maintained. Repairing of earthen dykes all around the bheri and sluice box, desilting of inside channel and mending of bamboo pattas and traps are done during the offseason. The farmers allow to the bheries to dry up and later plough the beds with the addition of lime, organic matter as well as inorganic fertilizers. Application of lime in bheries is done to disinfect the soil and for raising the pH, helping mineralization and faster decomposition of organic matter.

In saline bheries, tidal waters from rivers drawn into bheries during lunar period (new moon and full moon) for 4 to 55 days through the sluice with the shutter kept open. After filling the bheri, the excess water is let out through the sluice with shutter down during neap tide. During fish growing period the average water depth of about 1 meter is maintained.

In wastewater bheries sewage is drawn directly from the corporation channel. This water is highly polluted but rich in nutrients and organic matter. The sewage is drawn into bheri to a depth of about 80 to 90 cm and allowed for settling its suspended organic solids and oxidation of effluents by holding for about 20 to 30 days. When water becomes clear, then the fish seed is stocked. Further quantity of sewage is introduced at suitable intervals in a regulated way so that the environment is not degraded to affect fish growth and production.

Plankton forms the natural food in bheires. Water and soil are monitored regularly to check natural food status. When there is a deficiency of natural food, the supplementary feeding is done.

Generally the chemical fertilizers and artificial feeds are not used in bheries. The entire production system is based on exploitation of natural fertility of bheries. However few farmers use organic manures, oil cakes (mustard and mahua) and chemical fertilizers like urea and super phosphate for fertilization. Oil cake and rice bran are used as supplementary feed.

Before stocking in bheries fish and prawn post larvae are reared in small ponds. It is because they are very weak to survive in large waterbodies. The lack of adequate food and presence of carnivorous fishes make their survival poor. Hence, nursery management is essential to achieve high production in bheries. After nursery rearing when the fishes and prawn seeds have grown to young ones and also become hardy, are stocked in bheries.

The stocking density depends on the availability of fry and fingerlings of fish. Indian major carps at the rate of 50,000 to 60,000/ha, exotic carps (silver carp and common carp) at the rate of 3000 to 5000/ha and tilapia at the rate of 1000 to 1500/ha is generally stocked. The size of seed varied from 25-80mm.The stocking of fry is done from February to September. The harvesting of fish is done from April to November. The fish yield ranges from 2500 to 4000 kg/ha.

In saline bheries tidal waters from river are drawn into bheries initially to a depth of water about 20-30cm during December-January and is left as such for about 15 days for the growth of natural food. Immediately after that, the water level is raised and prawn and fishes are stocked. Low and medium saline bheries where in both freshwater and brakishwater fishes and prawn are grown have more or less same pattern of stocking. Prawn is stocked at the rate of 40,000 to 50,000/ha. After about a month, parse, major carps and tilapia are stocked at the rate of 1000-1500, 4000-5000 and 5000-6000/ha respectively in low saline bheries. In medium saline bheries, major carps fingerlings are stocked in monsoon months when water salinity drops. The production varies from 400-800 kg/ha.

In high saline bheries, besides autostocking of seeds along with tidal waters, selective stocking is also followed. Prawn, parse and bhangan are stocked at the rate of 30,000 to 40,000/ha, 700-1000/ha and 500-800/ha respectively. Tilapia is also stocked when salinity drops. These are of perennial type of fish culture with continuous trapping growing and harvesting practice. The fish and prawn yield ranges from 500-1500 kg/ha.

During the culture, farmers face several problems like prawn mortality, algal fouling, red disease, white spot disease, soft shelling etc. The different types of fish diseases such as dropsy, finrot, whirling and argulus infection are commonly recorded from the bheries. When disease strikes, usually farmers prefer to sell the produce if it is of harvestable size or even below that. During disease outbreak they resort to application of lime, that to a lower dose.

Bulk of prawn is harvested from April to July and other fishes from April to November. Prawn harvesting from bheries is quite interesting. Farmer erects 'bajas' (made up of bamboo sticks) inside the ponds, in which the prawns are usually trapped. For harvest of other fishes, gill net, cast net and drag net are operated.

Most of the farmers sell their harvested prawn and fish in the local 'arat'. It is a common selling place. Some farmers sell their produce to prawn processing companies. Marketing is not free from complications because of the involvement of middlemen and local dalals. For this reason farmers sometimes loose money.

*Chapter 3*

# Paddy-cum-Fish Culture

The culture of fish in paddy fields is of great significance in the economy of rural areas. It can provide a supply of cheap and wholesome protein food, besides affording an additional income. In the global level, this system was prevalent in as many as 107 countries, covering 135 million hectare of paddy plots during 1971. But in 1975, only 44 countries continued to have the system. After suspension of trawl fisheries, Japan had to take up rice-based fish culture during II$^{nd}$ world war. During mid-sixties in Malaysia, rice based fisheries contributed substantially to the freshwater production and in Indonesia a production of about 9 million kg of fish per year could be obtained from 90,000 hectares of arable land (Ghosh, 1990).

In India, there is a considerable scope for this practice in states of Assam, Bihar, West Bengal, and Orissa where enough water is present in the paddy fields, which otherwise left unused during a certain period of the year. However, in recent years, interest in paddy-cum-fish culture has declined because of indiscriminate use of pesticides to protect high yielding varieties of paddy. These pesticides if used properly will in no way harm the fish population around the paddy fields.

## Selection of Paddy Plot

Soil type is an important criterion for selection of paddy plot. Almost every type of soil like sandy loam to clay is suitable for paddy growing but clay soil preferred for the better paddy production and its high water retention capacity, ground water level, and drainage are the other features to be considered before construction of plot. Therefore, the proposed plot having uniform contour and high water retention capacity is preferred.

## Types of Paddy Plot to Integrate Fish Culture

The renovation of paddy plot depends according to the land contour and topography.

### Perimeter Type

Paddy growing area is in the middle with moderate elevation and ground sloping on all sides into perimeter trenches.

### Central Pond Type

Paddy growing area is on the perimeter with sloping towards the middle.

### Lateral Trench System

Trenches are provided on one side or both sides of the moderately sloping field.

## Preparation of Paddy Plot

The plot is prepared by sloughing the land at least twice during months of April–May before sowing the seedlings. Raw cattle dung @ 3000 to 5000 kg/ha is applied to paddy grown areas before ploughing. Direct sowing of the desired variety seedlings is wise, especially in the low-lying areas after first shower of monsoon. As the monsoon advances, the entire paddy plot and trenches became a single sheet of water. The kharif crop is harvested during months of November/December. The experiments were conducted by Central Inland Fisheries Institute (CIFRI), at Rahara Research Center and obtained 1.0 to 1.5 ton production of kharif paddy.

## Fish Culture

Before sowing the kharif crop in the paddy plot, the unwanted grasses/plants should be completely removed from the paddy plot and its surrounding. To remove the trash and weed fishes, Mahua Oil Cake (MOC) at the rate of 250 ppm should be applied.The organic fertilizer, which was applied during the preparation of paddy field, should also be applied in waterways at the same rate along with the inorganic fertilizers. This will help in producing the planktons and benthos.

## Fish Stocking

The fish species which could be cultured in rice fields must be capable of tolerating shallow water (<15 cm), high temperatures (up to 35°C), low dissolved oxygen and high turbidity. Fish seed is generally stocked during June to August. Indian Major Carp fingerlings *viz. Catla, Rohu* and *Mrigal* can be stocked at the rate of 5000 to 6000 nos/ha. Fishes like *Labeo bata, Puntius javanicus, Chanos chanos, Oreochromis mossambicus, Anabus testidineus, Mugil spp, Channa striatus* and *C.marulius* have also been found to respond well when cultured in the perimeter canal system. Culture of *Hypopthalmichthys molitrix* (Silver carp) and *Clarias batrachus* on the perimeter canal is also profitable.

To increase the fertility of the trenches or canals during December to March, cowdung (5000 kg/ha), Ammonium sulphate (70 kg/ha) and Single Super Phosphate (50 kg/ha) may be applied in three equal doses. If available the domestic sewage water can also be applied for fish culture.

## Supplementary Feeding

Supplementary food is given to enhance the fish growth rate. Groundnut oil cake and rice bran in the ration of 1: 1 may be given to the fishes at the rate of 2.5 per cent of their body weight particularly after the harvest of khariff crop.

## Fish Production

In this practice, the fishes are being cultured for a period of 5-6 months in the entire area and 4-5 months in the trenches. The fish production ranges from 500-700 kg/ha/year as obtained by CIFRI. Devaraj and Natarajan (1973) stocked 50, 900 fingerlings of different cultivable fishes in 11.9 hectare of paddy field and recorded production of 240-kg/ha area in 9 months period. *Cyprinus carpio* grow fast in paddy fields giving an average yield of 6.7 kg/acre, provided that the measures are being taken to prevent high rate of their mortality. Among the other carps the rohu and catla have better survival rate but poor growth in comparison to the common carp (Anonymus, 1962). The average yield of tilapia is reported to about 31.3-kg/acre area of the field (Anonymous, 1959).

*Clarias batrachus* is in great demand in the north–eastern parts of India, particularly in West Bengal, Assam, Orissa and Bihar for its high nutritional value. It contains higher percentage of protein and iron as compared to other edible fishes. Its fat content is also very low and is therefore easily digestible so that it is very useful during convalescence. With most other it is a delicacy because of the characteristic aroma and softness of its flesh. Magur is available in plenty, if not abundantly during the post-harvesting period of 'Aman paddies' in the sundarbans and other low lying areas of West Bengal.Magur is more profitable than that of any other cultivable species of fish. It is an ideal fish for culture in the paddy fields. Due to their omnivorous nature, they can consume a wide variety of insects as food. Moreover, they are hardy, easy to handle and economical to maintain. On the other hand, the high demand of magur, its fast growing nature, short culture period and use of paddy fields for rearing of fishes with selective and limited use of pesticides will commercially make the new system of integrated farming a success.

Besides this single tier system of integration of fish culture with other crops, there are other systems also where this integration may be 2 tier (fish-duck-horticulture) or 3 tier or 4 tier, depending upon the number of crops integrated. In China, a complex integrated farming is practiced in community ponds where live stock-piggery-duckery-poultry-floriculture-horticulture and agriculture are integrated with fish culture in large sized ponds to obtain maximum yield. But this type of complex integrated system requires efficient water and crop management skills; otherwise the whole system may be collapsed, because of high organic load in the ecosystem.

Though this system is not uncommon in China, but in India the simple tier integrated fish farming is even not popular and is yet to be started on scientific commercial lines.

## Control of Paddy Through Biological and Chemical Means

With the introduction of high yielding varieties of paddy in different parts of India, the farmers have started cultivating paddy in different seasons of the year from the same plot. But these high yielding varieties are prone to insect pests, such as rice hispa, stem borers, rice grasshoppers, brown plant hoppers etc. which are found to affect leaves, shoots and stem of paddy plants. The control of insect pests is very important as that of production management in paddy-cum-fish culture. The indiscriminate use of pesticides leads to the destruction of useful non-target species like birds, fishes etc. Some of the pesticides leaves toxic residues in crops, soil and fishes and poses a serious threat to human health.

## Advantages in Paddy-cum-Fish Culture

1. Economical utilization of land,
2. Little extra labour
3. Savings on labour cost towards weeding and supplemental feeding,
4. Enhanced rice yield, and
5. Additional income and diversified harvest such as fish and rice from water, and onion and sweet potato through cultivation on bunds.

## Problems in Fish Culture in Paddy Fields

Most common problem faced in paddy-cum-fish culture is the loss of fish during the period of their growth. The loss amounts to about 40 to 60 per cent for the young and 20 to 30 per cent for the large fishes. It is due to birds like herons and others etc. living on the fish. Inadequate oxygen level, abrupt change in temperature and the poor depth of water also contribute significantly to the loss of fish.

## Remarks

Paddy-cum-fish culture makes multiple use of the paddy field to maximize the utilization of land and water resources and can also increase the production value of paddy fields. Fish as a bio-controller of insects and weed, reduces the input cost of insecticide and herbicide and toxicity accumulation is minimized. This is beneficial to human health and the ecological balance of the environment.

## REFERENCES

Ghosh, Apurba, 1990. Rice-cum-fish culture and its economic feasibility. In: *Technologies for Inland Fisheries Development* (Eds.) V.V. Sugunan and Utpal Bhaumik. Central Inland Fisheries Research Institute, Barrackpore, pp. 69–74.

Mohanty, Rajeeb, K., 2006. Eco-agriculture through integrated paddy-fish farming: A step towards productivity maximisation. *Fishing Chimes*, 25(12): 17–20.

*Chapter 4*

# Fish-cum-Duck Farming

## Introduction

Fish-cum-duck farming is an economically viable and technologically simple system and thus can be taken up by the rural farmers to improve their economic condition. In the system the land area in the farm *viz.*, embankment, dykes etc. can be used for constructing duck pens, feeding ground for ducks as these areas harbor insects, worms, etc. Duck dropping can be used effectively for manuring pond and by rearing proportionate number of ducks; the necessity of manuring and supplementary feeding can be eliminated. Also ducks feed on aquatic insects, tadpoles, molluscs, snails, soft aquatic weed (lemna, azolla etc.) which are harmful to fish culture and thus help in keeping the environment clean. Again, in their search for food, ducks dabble the pond bottom, which helps in releasing obnoxious gases as well as nutrients from the pond bottom. Thus, in the system everything is effectively utilized for the production of valuable proteinecious crops. The expenditure spent on duck farming is largely offset by the sale proceeds of the duck flesh and eggs. In this system of farming about 3, 500 to 4,000 kilogram of fish and 500 to 600 kilograms of duck meat and 18000 to18,500 eggs can be obtained from a hectare pond area in a year.

Fish-cum-duck farming is practiced in Tamil Nadu, Assam, Bihar, Andhra Pradesh, Tripura, Orissa, Karnataka, Kerala and Uttar Pradesh.

## Selection of Ducks

In selecting ducks two things should be borne in mind; sturdiness so as to withstand the weather conditions of the open pond in all seasons and the egg laying capacity so as to earn more income. Two types Indian runner and Khaki cambel are found to be most suitable for this system of farming. The egg laying capacity of Indian runner is 150 eggs in a year and that of khaki cambel is 180 to 200 eggs in a year and their growth rate is about 1 kilogram in a year.

## Rearing Density

The object of duck raising in this system is to get duck easily to manure pond water under fish culture and thus the number of ducks should be adjusted so as to get adequate duck dropping to manure the pond. It has been found that 500 to 1500 ducks are sufficient to manure the pond of one hectare water area under composite fish farming with their dropping.Two to three week old ducklings should be introduced in the system after administering necessary prophylactic medicines to avoid mortality and much expenditure.

## Feeding

Ducks are set free in the farm during the daytime and thus they are able to find their natural food such as insects, worms, tadpoles, molluscs, soft aquatic weeds, etc. from the pond water and the embankments. But this is not sufficient for their proper growth and hence artificial food may be given to supplement the natural food, by mixing any standard balanced poultry feed with good quantity rice bran in the ratio of 1: 2 by weight. The feeding of this mixture may be done at the rate of 100 gm per each duck in a day in two times in equal halves-first one in the morning and another one at the evening either in the duck house or on the pond embankment so that the spilled feed can be easily drained into the pond. The feed should be mixed with enough water in the container so that the bill of the duck can be submerged at the time of tasking feed as the duck can not feed properly without water.

## Egg Laying

Ducks start laying eggs at the age of 24 weeks and continue to lay egg for two years. Generally the duck lays eggs at night. Each duck lays about 150 eggs, 180-200 eggs (khaki cambel) in one year.

## Health Care

An experienced farmer can detect a sick bird among the flock by a careful look, listening to the sound of the bird and by observing any reduction in daily feed consumption. A sick bird becomes listless, eyes lack brightness and watery discharge comes out of eyes and nostrils. The ducks are to be vaccinated prior to stocking.

## Harvesting

Harvesting should be done after two year rearing since after two years; the egg laying and fattening capacities are reduced considerably. It is rather more economical to dispose them off after one year rearing as they cost more than that age due to softness of their flesh. Egg collection can be done in the morning every day after lefting ducks out from the pen. It is expected generally to obtain 500 to 1500 kilograms of duck meat and 36,000 to 75,000 eggs in one-year rearing period.

## Use of Duck Dropping as Manure

The duck droppings contain 25 per cent organic and 20 per cent inorganic substances, and are rich in carbon, nitrogen, phosphorus, potassium, calcium etc. Ducks are called as 'living manuring machines' and provide very good fertilizer to the fish pond and dabbling of ducks at the bottom in search of food releases nutrients from the soil, further enhancing production of pond. They feed on insects, larvae, tadpoles, molluscs, worms and aquatic weeds, which do not form food of stocked fish, and thus the two are highly compatible. Ducks are freely allowed to stay in the pond during the day so that their dropping fall automatically all over the pond. Dropping voided at night in the duck pens are washed off and the washing is drained into the pond every morning.

A duck voids on an average 125 to 150 gm droppings every 24 hours and thus 200 to 300 ducks would void on an average 10,000 to 15,000 kg of droppings during the year, which is sufficient to manure one-hectare water area under composite fish culture. The duck dropping has high manuring value.

## REFERENCES

Sharma, B.K., Das, M.K., Das, S.R. and Neogi, S.K., 1979. Observations on fish-cum-duck rearing. Symposium on Inland Aquaculture, Barrackpore, India (Abstract), p. 95

Sharma, B.K. and Das, M.K., 1988. Studies on integrated fish-livestock crop farming system. In: Joseph, M. Mohan (Ed.), *Proceeding of the First Indian Fisheries Forum*, Asian Fisheries Society, Indian Branch, Mangalore, Karnataka: 27–30.

Ziauddin, Golam and Arunima Goswami, 2005. On integrated fish farming. *Fishing Chimes*, 24(11): 13–16.

*Chapter 5*

# Bundh Breeding of Fishes

The main problems in riverine spawn collections are dependence of flood positions, admixiture of both desired and undesired varieties of fish seed, location and approach to suitable collection grounds, transport facilities etc. To circumvent the problem of availability of quality fish seed, the fishermen of West Bengal established a special type of fishery operation called 'Bundh breeding' in early 1900s in the districts of Midnapur, Bankura and Chittagong (Now in Bangladesh). Bundh is a local term that means a dyke or an embankment. However, with regard to fish breeding, bundhs are special type of perennial or seasonal ponds or impoundments where riverine conditions are stimulated during the monsoon period owing to the accumulation of rainwater from the catchment areas. These are two types, *i.e.*,

1. Perennial or Wet bundhs; and
2. Seasonal or Dry bundhs

## Dry Bundh

It is a shallow seasonal pond or a depression on the land, bounded by embankments on three /four sides which impound freshwater from the catchment area during the monsoon season. It remains dry for a considerable part of the year. In West Bengal, a bundh having a catchment area of more than 5 times the area of the bundh proper is considered most suitable. But in Madhya Pradesh a ratio of 1: 25 between the area of the bundh and catchment area is considered essential. This ratio depends to a large extent on the average rainfall, in the catchment area. Dry bundhs in Madhya Pradesh are relatively bigger in size (0.2 to 2.5 ha) than those in West Bengal (0.1 to 0.5 ha).

Initially sufficient quantity of rainwater is stored in the dry bundh and brood fishes are introduced into it, from a perennial pond. Spawning is generally observed during and after heavy showers. After breeding is over, the bundhs are partially drained and the eggs are collected for hatchling.

In West Bengal, a seasonal pond is selected in an undulating terrain to receive and store rainwater from the adjoining catchment area and is known as a 'reservoir' or 'storage tank'. Small ponds of 0.1 to 0.5 ha with sloping bottom, having inlet and outlet, are constructed at a lower level than the reservoir. These small ponds or bundhs are connected with the reservoir by channels to have a constant flow of water throughout the period of breeding by gravitational force. The actual number of bundhs depends on the total storage capacity of the reservoir. Both the reservoirs and bundhs are seasonally operated.

In modified type, before the breeding programme the inlet is opened and water is allowed to accumulate up to 30-60 cm. The mature males and gravid females (1: 1) are selected and introduced into dry bundh. The brooders are left undisturbed for 6-12 hours to get acclimatized to the new environment. Then intramuscularly injection of fish pituitary extract @ 5-10 mg/kg bodyweight of fishes are administered in a single dose to 10 to 20 per cent of the stocked males as well as females. Females are generally injected with a little higher dose than the males in evening or night hours. Now the inlet and outlets are opened so that rainwater from the reservoir gushes into the bundh and goes out. Proper attention is essential to guard the inlet and outlet with iron meshed screen to check the escape of brooders. The regular water flow is maintained till the sex play starts (sex play starts after 4 to 6 hours of injection). The flow of water is regulated and finally both the inlet and outlet are closed. The breeding continues for 4 to 6 hours. Mostly all the fishes released in the bundh participate in the breeding activity irrespective of whether one received the pituitary extract dose or not. This phenomenon is known as 'Sympathetic breeding'.

After breeding, the eggs are left undisturbed till the embryo is fully formed. Two persons collect the eggs after lowering the water level by dragging a mosquito netting cloth piece. After that, eggs are transferred for hatching to the small earthen pits with mud-plastered walls. These pits are called as 'chaba' and approximately 3'x2'x1' in size. Usually 2 to 3 liters of fertilized eggs (50,000 to 75,000 in numbers) are dumped in each pit and water from the reservoir is circulated through the pits. Hatching of eggs takes place 15-20 hours after fertilization. The hatchlings are collected from the 'chabas' with help of muslin cloth pieces and transferred to similarly prepared bigger sized earthen pits (4'x3'x$1^{1/2''}$) locally called as 'hamar'. A total number of 3 to 5 lakh hatchlings are usually dumped in each hamar. The selling starts 48 hours after hatchling or even earlier.

In Madhya Pradesh, sufficient rainwater is accumulated in dry bundh before the broodfishes are introduced. The fish farmers generally prefer a rainy day for this purpose. The hatchling of eggs is processed in double walled cloth hapas or in cement cisterns.

## Wet Bundhs

It is perennial pond situated in the slope of a vast catchment area. Wet bundhs are comparatively much bigger in size than the dry bundhs. They have proper embankments equipped with inlets at the up-land side and outlets towards the low-land zone. During heavy rain, a greater portion of the bundh becomes inundated. The excess water flows out through the outlet. A wet bundh can be of any shape and dimension. Normally, the catchment area of 20 to 100 times the area of the bundh is considered better for breeding. With the onset of monsoon, the rainwater from the catchment area gushes to the bundh where the brood fishes are already present.In the wet bundh when the conditions are favorable, the fishes start breeding and the eggs are collected and hatched in the same way as in dry bundh. Construction of wet bundh is extremely expensive and not a profitable proposition. However, the existing bundhs can be developed and improved for carp seed production.

Several factors are responsible for spawning of carps in bundhs. Spawning is stimulated by heavy monsoons that flood the shallow spawning areas. It appears that availability of shallow spawning grounds inundated with fresh rainwater is an important factor in stimulating the brooders. According to some workers, spawning occurs in still waters, while others believe that strong current on the spawning grounds favours breeding. The favorable temperature of water for spawning has been found varies from 24 to 32°C under various environmental conditions. Generally the cloudy days followed by thunderstorm and rain, is regarded to influence spawning. Other factors like pH, alkalinity, high oxygen content etc; cannot be ignored. These factors are associated with the floods. Generally, the bundhs having highly turbid waters with a distinct red colour, low pH between 6.2 to 7.6, 5 to 8 ppm of dissolved oxygen, low total alkalinity and 27 to 29°C temperature provide favorable conditions for spawning in bundhs.

## Economics of Bundh Breeding

In Midnapore and Bankura (West Bengal), 75 lakhs of spawn was produced at a time, and 160-220 million spawn produced in a season. With the increasing pace in the creation of a large number of bundhs, it is necessary to mention that spawn production through dry bundhs, is quite economical. Many crops of seed can be easily obtained from one bundh in a season of four months.By utilizing the rainwater, which would otherwise have been wastewater; we can produce carp seed and reap good profits.

## Problems in Bundh Breeding

1. The wet bundh being larger and deeper has operational difficulties
2. Percentage of the collection of eggs is poor due to large size of breeding area
3. Leftover eggs perish due to rapid depletion in oxygen
4. Eggs and hatchlings are preyed upon by the predatory and weed fishes
5. The seed production in wet bundhs is not so pure, because it harbours some weed and predatory fishes which breed along with carps.

*Chapter 6*

# Role of Soil and Water Quality in Fish Culture

Water and bottom soil are directly concerned in the culture practices of fish. Physical, chemical and biological properties of soils, denuded and transported from surrounding areas and deposited in the basin shaped areas, vary from those of the cultivated soil from where it has been removed. A number of soil constituents such as colloidal clay, organic matter, nitrogen, phosphorus, potassium etc and also trace elements are carried down with flowing water, get distributed in different properties at various parts of fish ponds which ultimately determines the physico-chemical and biological characteristics of the water of fish pond.

## Soil

## Physical Properties of Soil

Texture, structure, bulk density and colour are the main physical properties of soil. An ideal pond soil should have sufficient amount (50-60 per cent) of clay and silt in association with other soil particles. Fishponds with gravel and sandy bottom are unsuitable for fish culture. Soil structure with loamy or clay loam bottom and dark in colour is categorized as most productive fishpond.

## Chemical Properties of Soil

Soil reaction, organic carbon, available nitrogen and available phosphorus are the main chemical features to categorize the fish pond as low, intermediate or high productive in nature.

### pH

Soil reaction is the main factor, as it determines the acidic, neutral and alkaline condition of soil. Most of the essential plant nutrients are released in balanced form under neutral condition of soil pH *i.e.,* between 6.5 to 7.5. Highly acidic soil is unsuitable for fish culture. Under acidic soil, loss of appetite takes place among the fishes. Highly alkaline condition of soil creates destruction of soil structure and also retards the fish growth. Microbial activity is also affected by the soil pH.

### Organic Carbon

Pond bottom soil with very low content of organic carbon is not suitable for fish culture due to its low productivity in nature. Soil over 1 per cent organic carbon is ideal for fish production. Very high organic content is undesirable because it will result in high C: N ratio. Under high C: N ratio *i.e.,* over 10: 1 will adversely affect the microbial activity and most of the soil nitrogen will be utilized by micro-organisms and soil will loose its nitrogen supplying capacity.

### Nitrogen

Soils may be having rich source of nitrogen. But depending on the other limiting factorts, if release of available nitrogen is restricted, the soil is rendered unsuitable for establishing fishpond on such soil base. Soils with available nitrogen above 50 mg N/100 g soil is considered to be suitable for fish production. Availability of soil nitrogen is mainly affected by microbial population of the soil in association with other environmental factors.

### Phosphorus

It is present in various forms in the soil. Important forms are organic (proteinous and nonproteinous) and inorganic (calcium, phosphate, iron, aluminuim and pyrophosphates). Availability and release of soil phosphorus is mainly pH dependent. Under alkaline condition, the available forms of soil phosphorus are converted into octocalcium form and hydroxy appetite is formed and under highly acidic condition, insoluble form of iron and aluminium phosphate is formed and phosphorus becomes unavailable to the organisms. Available phosphorus above 10 mg P/100 g soil is favorable for fish culture.

## Biological Properties of Soil

Microorganisms play an important role in determining the physico-chemical characteristics of soil. Soil microbes are highly sensitive to the soil reaction. Soil pH at and around neutral is favourable for healthy growth of soil micro-organisms and microbial activity. Soil fertility based on microbial activity was studied in bottom, side and adjoining areas of small reservoirs. Atmosphere is rich source of nitrogen (more than 79 per cent) but it cannot be utilized by land crops, until and unless it is somehow trapped by the microbial activity in the soil. Several forms of microorganisms are present in soil and are of great importance as they help in the decomposition of organic wastes and release of nutrients. Nitrogen cycle is the best example for bacterial activity in the soil.

# Water

## Physical Properties of Water

Temperature, colour, transparency, turbidity are among the important physical features of pond water.

### Temperature

Temperature, very low or very high are not suitable for fish culture though fish can withstand a wide range of water temperature. Water temperature between 15 and 35°C is favorable for healthy growth of fishes. However, for breeding purposes, temperature between 20 and 30°C is ideal.

### Colour

Colour of water should be indicative of plankton bloom, throughout the culture period.

### Transparency and Turbidity

The transparency of water *i.e.*, the least level of turbidity in water enables the penetration of light to the depths where it is ultimately absorbed.In deep water or reservoir water, the zone up to which the light rays penetrate is called as 'photic zone' and below this, there is a complete darkness and the organisms that require light (*e.g.* algae) cannot live. Sometimes manifestation of phytoplankton due to the application of organic manures in excess adversely affects the productivity of pond. Photosynthetic activities of plants in turbid water get reduced and the fish fauna needing clear oxygen-rich water are replaced by predominantly labyrinthiform fauna in such a situation. Behavioural changes in fish have been noted at clay turbidities above 20,000 mg/l and appreciable mortality has been noted above 1, 75,000 mg/l. Clear ponds having clay turbidity within 25 mg/l have been seen to be more productive than those having turbidities above 25 mg/l (Buck, 1956). Thus, productivity and photosynthesis seem to be inversely proportional to turbidity. Accumulation of humic substances is also harmful for fishery because of restricted light penetration and development of acidity in such waters. Plankton turbidity is generally good for fish production but this also becomes a menace during eutrophication caused by a high planktonic bloom (Boyd, 1982).

## pH

The pH is affected not only by the reaction of $CO_2$ but also by the organic and inorganic solutes present in water. Any alteration in water pH is accompanied by the changes in other physio-chemical parameters. The pH of a water media often gives indication of fertility or potential productivity in a pond. The diurnal fluctuation of pH of a water body should remain in the range of 6.4 to 8.5 in order to support the optimum fish growth. If waters are more acidic than pH 6.5 or more alkaline than pH 9-9.5 for long periods, reproduction and growth have been found to be affected (Swingle, 1961; Mount, 1973). The pH range of 6 to 9 is most suitable for pond fish culture. While pH more than 9 is unsuitable (Swingle, 1967). At pH 4 or 11, fish will die out with the development of ulcers as a result of severe corrosion of fish body (observation of R.K. Das at Domjur, Howrah, West Bengal, 1993).

## Dissolved Oxygen

Dissolved oxygen is a vital parameter as it is directly involved in the metabolic activity of flora and fauna of fishpond. The main sources of oxygen in the pond water are (*i*) diffusion of atmospheric oxygen and (*ii*) photosynthetic production by phytoplankton. During the process of photosynthesis, carbon dioxide is reduced to carbohydrate by the chlorophyll bearing organisms and oxygen is liberated. Solar energy is used and stored in the form of chemical energy. Oxygen thus produced during the process of photosynthesis is utilized by other organisms and equilibrium is maintained.

The dissolved oxygen content of warm water fish habitats should not be less than 5 mg/l, during at least 16 hours of any 24-hour period. It may be less than 5 mg/l for a period of not exceeding 8 hours within any 24-hour period and at no time shall the dissolved oxygen content be less than 3 mg/l (Boyd, 1982; Mc Kee and Wolf, 1963).

Continued exposure to low dissolved oxygen is considered as a precursor to bacterial infection in fish. Exposure of fish to dissolved oxygen less than 1mg/l is lethal when it lasts longer than a few hours (Swingle, 1969), because the minimum concentration of dissolved oxygen tolerated by fish is a function of the exposure time. Further, the minimum tolerable concentration of dissolved oxygen will vary with species, size, physiological condition, concentration of solutes, temperature and other factors. It has been found that fish can survive at dissolved oxygen in the range of 1 to 5 mg, but its growth and reproduction are retarded.

Sugunan (1989) mentioned the dissolved oxygen below 5 ppm can be lethal to the biota. So high dissolved oxygen content of pond water is desirable and important factor in productivity.

**Carbon Dioxide**

Natural waters generally accumulate carbon dioxide through respiration of aquatic plants and animals. In low concentration it indicates productiveness of a water body (15mg/l) but above 15 mg/l, it hinders the uptake of oxygen by fish (Boyd, 1982). Accumulation of dissolved carbon dioxide also increases with dying of phytoplankton and with cloudy weather when photosynthesis becomes limited. At higher concentration of $CO_2$, pH decreases and water becomes acidic both of which are not conductive for fish health and growth. At night, with decrease of dissolved oxygen, concentration of dissolved carbon dioxide increases and this makes the loading of oxygen more difficult for fish (Boyd, 1979). Thus, at lower concentration $CO_2$ is good for photosynthesis and at higher concentration it is harmful for fish growth and production.

According to Agarwal (1999) the carbon dioxide content of the water depends upon the temperature of water, depth of water, rate of respiration, decomposition of organic matter, chemical nature of the bottom and the geographical and physiological features of the terrain surrounding the water. The carbon dioxide content of the water increase with increase in temperatures and is generally higher in tropical waters than in temperate or polar waters.The vertical distribution of carbo dioxide in water column has inverse relationship with oxygen levels. Both free and bound carbo dioxide are relatively low at the water surface and increases with water depth and are plentiful at or near the bottom.

Lagler (1978) mentioned that the amount of free carbon dioxide in water is important in fish management because it is perhaps the best single criterion of environmental suitability for fishes. High concentrations of free carbon dioxide which are in themselves toxic to fish are usually accompanied by low values for dissolved oxygen. Free carbon dioxide in excess of 20ppm may be regarded as harmful to fishes, although lower values may be equally harmful in waters of low oxygen content (less than 3 to 5 ppm).

**Alkalinity**

Alkalinities refers to the total concentration of bases in water expressed in mg/l of calcium carbonate. In most waters bicarbonates, carbonates or both are predominant bases. The alkalinity determines availability of free $CO_2$, which is essential for photosynthesis and thus is directly linked to productivity (Wallen, 1955). In general alkaline water is more productive and supports diversity of aquatic life; the desirable range being 100 to 125 mg/l (Huet, 1948). Waters having alkalinities below 20mg/l, show sharp rise of fall of pH with small change in the concentration of bases or acids because of poor buffering capacity of such waters. Waters with such poor buffering capacities often becomes unconducive for fish production. Waters having 40 mg/l or more total alkalinities are considered to be more productive than waters of lower alkalinities (Moyle, 1945; Mairs, 1966). It has also been observed that the outbreak of Epizootic Ulcerative Syndrome (EUS) was more severe with waters having alkalinities below 20mg/l (Das and Das, 1995).

**Hardness**

Water hardness is, from a fish's viewpoint, one of the most important aspects of water quality as it affects so many areas of fish health. Water hardness has a major effect on pH and pH stability. It has a major effect of fish osmoregulation. Different species of fish have varied water hardness requirements. The hardness of water is mainly due to the presence of calcium and magnesium. According to Jhingran (1988) and Sugunan (1990) the hardness above 70 ppm is a indicator of the better productivity. Fishes have been found to be susceptible to diseases in water with hardness below 20mg/l. Productive waters should have hardness above 20mg/l, calcium above 5 mg/l and magnesium above 2mg/l. Very hard water (>300 mg/l) also becomes uncongenial for fish production because of higher pH. Optimum hardness for fish culture has been observed to be around 75-150 mg/l.

**Nitrogen**

It is one of the basic elements of fertility in ponds and is present in water in an amount of 65 per cent of all dissolved gases at equilibrium. Certain quantities of nitrogen may reach soil and water from the atmospheric nitrogen by rain or incidental lightening. There is also role of soil bacteria that can fix atmospheric nitrogen and, thus, make it available to the green plants. Whatsoever, are the alternative possibilities, the ponds have the major supply of nitrogen from the decaying organic matters *i.e.*, from the putrefaction cycle in which nitrates are the end products. From the productivity of ponds, sufficient quantities of nitrogen compounds, such as nitrates are essentialities when water contains 4 ppm of nitrogen with 1ppm of phosphorus and 1 ppm of potassium the production of plankton is high.

**Hydrogen Sulphide**

The deeper layer of many ponds, tanks, lakes, and reservoirs may contain significant amounts of the toxic gas, hydrogen sulphides, which is formed by decaying organic matters. In this decaying action, anaerobic bacteria play the main role. Sometimes the gas accumulates in the fishponds having thick layer of organic deposits at the bottom. This gas can be a severe limiting factor in fishponds, as a concentration of 6 ppm of $H_2S$ will yield drastic effects for the aquatic biota in general and can kill common carp in few hours. In such a situation, aeration of ponds by mechanical agitation can help a lot to free the pond of hydrogen sulphide.

Unionized hydrogen sulphide is toxic to fish but its ions are not appreciably toxic. Adelman and Smith (1970) have shown that egg survival and fry development in northern pike (*Esok lucius*) are poor due to the presence of 0.06 mg/l of unionized hydrogen sulphide

**Phosphorus**

Phosphates and nitrates are essential for fish production in ponds. In all natural waters, these nutrients are found in varying quantities in dissolved form in the bottom soil. If nutrients are deficient, production and growth of organisms in the pond water are adversely affected. The phosphorus is released from the dead organisms in the bottom of the pond of giving rise to the growth of 1ppm of phosphorus. This level of phosphorus is considered as optimum for sufficient growth of plankton.

**Potassium**

This is present in a cultivable pond to an extent of 1ppm along with 1ppm phosphorus and 4ppm nitrogen. This concentration of the three constituents determines the fertility of the pond and is considered to be most ideal for the growth of planktons. Potassium is released from the bottom soil, which is rich in deposits of various kinds. It is released from there and is specially effective in stimulating the growth of aquatic vegetation.

## REFERENCES

Agarwal, S.C., 1999. *Limnology*. APH Publishing Corporation, New Delhi, pp. 150.

Boyd, C.E., 1979. *Water Quality in Warm Water Fish Ponds*. Auburn University, Agricultural Experiment Sation, Auburn, Albama, 359pp.

Boyd, C.E., 1986. *Water Quality Management for Pond Fish Culture.* Elsevier Publication, 318 pp.

Buck, D.H., 1956. Effects of turbidity on fish and fishing. *Trans. N. Amer. Wildi. Conf.*, 21: 249–261.

Das Manas Kumar and R.K. Das, 1995. Fish diseases in India: A review. *Environment and Ecology*, pp. 533–541.

Das, R.K., 1996. Monitoring of water quality, its importance in disease control. In: *Paper Presented in Nat. Workshop on Fish and Prawn Disease, Epizootics and Quarantine Adoption in India*. October 9, 1996, CIFRI, pp. 51–55.

Huet, M., 1948. pH values and reserves of alkalinity. *Wat. Poll. Abstr.* 21: 254.

Jhingran, A.G., 1988. Reservoir Fisheries in India. *J. of the Indian Fisheries Association*, 18: 261–273.

Lagler, Karl F., 1978. *Freshwater Fishery Biology*. WMC Brown Company Publishers, Dubuque, Iowa, pp. 234–259.

Mairs, D.F., 1966. A total alkalinity atlas for marine lake waters. *Limnol. Oceanogr.*, 11: 68–72.

Mc Kee, J.F. and Wolf, H.W–1963. *Water Quality Criteria*, 2nd edition. State of California, State Water Quality Control Board, Sacremento, publication-A, 548 pp.

Moyle, J.B., 1945. Some chemical factors influencing the distribution of aquatic plants in Minnesota. *Amer. Midl. Natur.*, 34: 402–420.

Sugunan, V.V., 1989. Salient features of reservoir limnology and their significance to fisheries development. In: Jhingran Arun G. and Sugunan V.V. (eds.). *Conservation and Management of Inland Capture Fisheries Resources of India*. Inland Fisheries Society of India, pp. 106–111.

Sugunan, V.V., 1990. Reservoir fisheries management. In: Sugunan, V.V. and U.Bhowmick (eds.). *Technologies for Inland Fisheries Development*. Central Inland Fisheries Institute, Barrackpore, pp 153–164.

Wallen, I.E., 1955. Some limnological consideration in the productivity of Oklahoma farm ponds. *J. Wildlife Mont.*, 19(4): 450–462.

*Chapter 7*

# Aquatic Weeds

Aquatic weeds are those plants, which start in water and grow at least part of their life cycle in water. In recent years there has been a growing awareness among fish culturists regarding aquatic weed problems. Presence of small amount of certain aquatic plants in fish pond may be useful at times as they have a definite role in the development and maintenance of a balanced community. But dense growths of weeds assimilate a large proportion of nutrients in the pond and thus compete with fish population, limiting available fish food. They restrict fish movement and interfere with fishing operations, besides serving as ideal hunts for predatory fishes and insects etc. The surface covers of floating weeds prevent light penetration. Excessive development of algae have also been causing nuisance in cultivable fish ponds and small tanks. They even cause fish mortality due to oxygen depletion or release of extra-cellular metabolites, which are toxic.

## Types of Aquatic Weeds

Based on habitate aquatic weeds are classified into five types *i.e.*, free floating, submersed, emerged, marshland, ditch bank, phreatophytes, and floating island weeds.

## Free Floating Weeds

These weeds have leaves that float on the surface water either singly or in rosettes. They may be free floating as in case of *Lemna* or

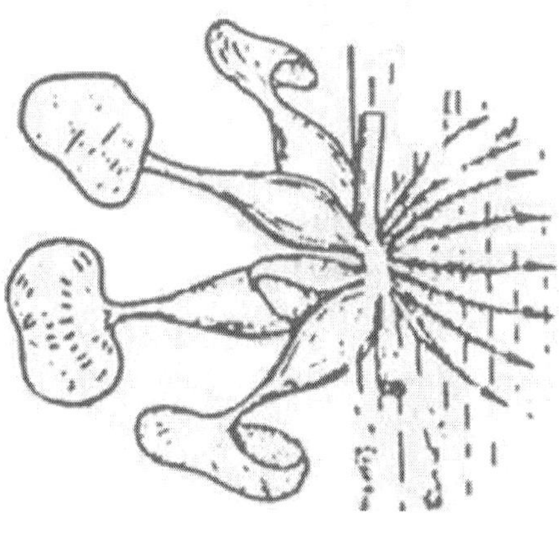

Figure 7.1: *Eichhornia*

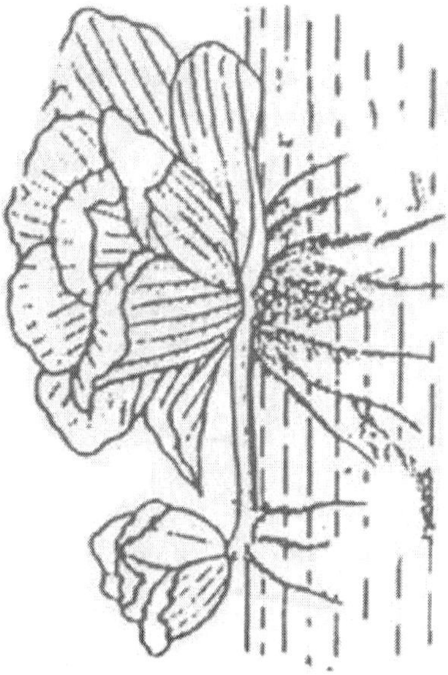

Figure 7.2: *Pistia*

may be enchored in the hydro-soil like *Nuphar lutea*. Some of them have submersed leaves, which are often of a different shape to those on the surface.

Surface floating weeds include *Eichhornia, Salvinea, Pistia* and *Lemna*.The anchored to the mud bottom weeds include *Nelumbo, Nymphea spp*. Some floating weeds like *Cyperus papyrus* may grow as massive mats.

## Submersed Weeds

These are mostly vascular plants which produce complete or partial of their vegetative growth under water. Some weeds send the upper leaves and inflorescence to the above water surface. These leaves are morphologically different than leaves that grow beneath the water surface.

Submersed weeds are of two types *i.e.*, free floating and anchored to the bottom mud. The floating weeds such as *Ceratophyllum spp*

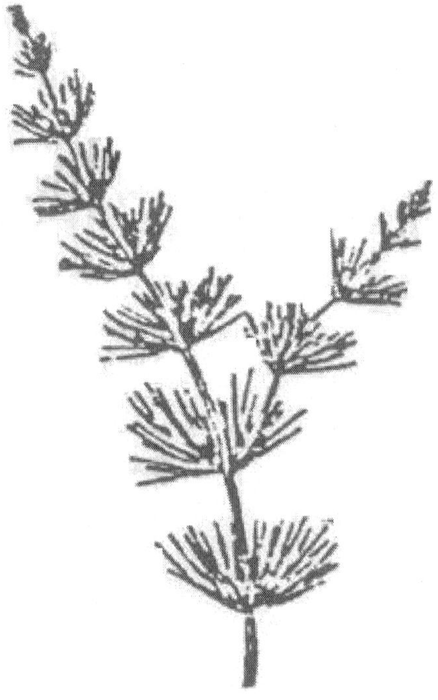

**Figure 7.3: *Ceratophyllum***

and *Utricularia spp* hang in water without any attachment with hydrosil. However, anchored species like *Vallisneria spiralis* strike roots in the bottom mud.

## Emerged Weeds

These weed have their roots beneath the water and their stems and leaves above the surface. These weeds grow in places where the water level ranges from just below ground level to about half of maximum height of the plant. The leaves of emerged plants do not raise and fall with water level as in case of rooted floating weeds.

The common emerged weeds are *Typha spp, Nymphaea spp, Sagittaria sp* etc.

## Marshland Weeds

These are very common in low-lying areas. When these lands flooded, they termed as marshes and when the free water is absent

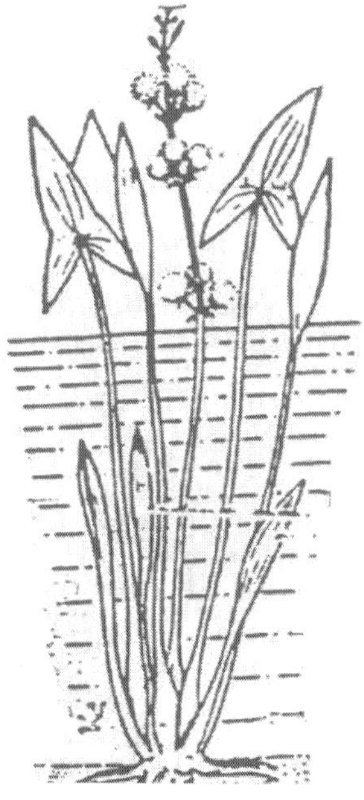

**Figure 7.4:** ***Sagittaria***

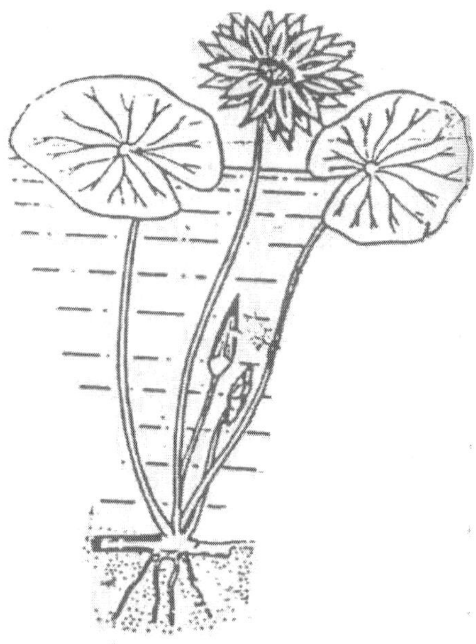

**Figure 7.5:** ***Nymphaea***

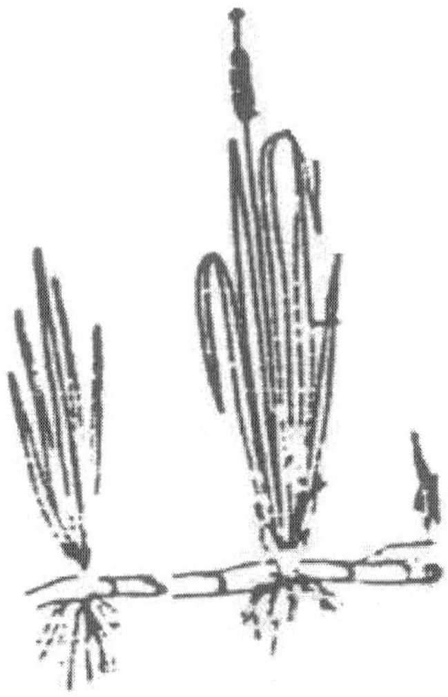

**Figure 7.6:** ***Typha***

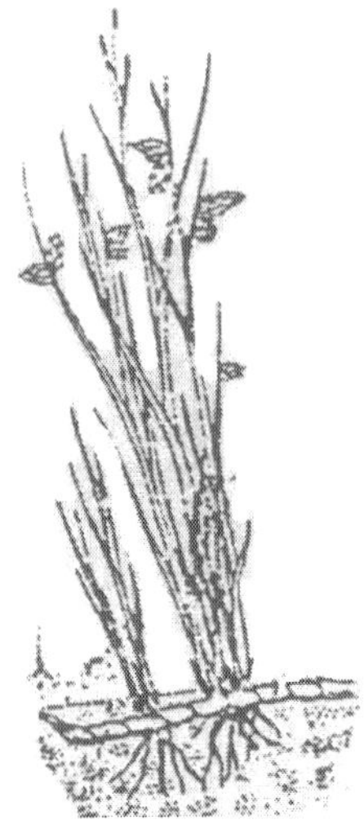

**Figure 7.7:** ***Scirpus***

these are known as swamps. These weeds are amphibious. The common marsh weeds include *Carex rostrata, Alisma plantago, Scirpus supinus* etc.

## Ditch Bank Weeds

These weeds grow on banks of drainage ditches and irrigation canals. They are land plants but prefer moist soils. Few species prefer higher moisture and others need dry lands. *Phragmites karka* and *Sorghum halepense* prefer high moist grounds. They germinate in moist soils or shallow submerged soils although they tolerate temporary flooding.

## Pherotophytes

These weeds send their roots down to the water table or the capillary fringe just above the water table, which provides a ready supply of water. These are the water loving plants, which grow along irrigation canals and other watercourses. Pherotophytes are most aggressive and usually invade shallow water table. The common phenotypes *Tamarix gallica, Salix spp, Sarcobatus vermiculatus* etc.

## Floating Island Weeds

These are large vascular plants that are grown vast masses of floating dead aquatic vegetation. These weeds are main problem in large lakes and swamps in African continent. It is very difficult to manage these weeds at the time of flash floods and create havoc in many areas.

## Methods of Control

The aquatic weed control methods may be broadly grouped under five different categories, *viz.*, preventive, manual, mechanized, chemical and biological.

### Preventive

Considering the high cost of clearance of weed infestations, certain preventive measures are suggested, which reduces the chance of weed incidence.

The preventive measures have to be thought of in advance and have to be integrated in the general management of water bodies. Some of the common measures considered are depending of pond margins, desilting of old silted ponds, ploughing or burning of marginal weeds and providing barriers to entry of floating weeds.

### Manual Method

Manual method is the age old and simplest method. It involves physical removal of weeds by manual labour. This is convenient and economical particularly in small ponds.

The free-floating weeds are hand picked or dragged ashore by wire or coir rope nets. The submerged weeds are either pulled by hand or hand drawn bottom rackes or uprooted with bamboo poles having a cross piece at their lower ends (Philipose, 1963). Repeated cutting of the leaves of water lilies as reported by Swingle and Smith (1950) is also useful at times. Hickling (1962) reported use of manual labour for repeated cutting of grasses and sedges of pond margins in Malaysia.

### Mechanical Method

Use of mechanical devices for removal of aquatic weeds have drawn considerable attention in recent years. The use of steel cables, cutting chains and diesel operated winches (Mitra, 1956) for clearance of rooted submerged weeds are prevalent. Cutting boats of different type having petrol operated underwater cutters are used in some regions.

Mechanical methods have certain drawbacks such as initial high cost, skilled operator for handling, difficulty in transporting and installing heavy machinery in remote areas.

## Chemical Method

This method is effective and long lasting in larger waterbodies. The weeds killed by chemicals are allowed to settle at the bottom and ultimately recycled to become manure for enhancing fertility of fishery waters. Sometime, however, the sudden death and decomposition of large masses of vegetation may deplete dissolved oxygen in water even causing fish kill. Hence, technical guidance is essential in regard to choice of the herbicide and methods of application. No one chemical has been developed so far which would control all aquatic weeds. So it is essential to know the weed species, appropiriate herbicide and their rate and time of treatment.

The common herbicides used for chemical control of aquatic weeds are depicted in Table 7. 1.

**Table 7.1: Common Herbicides Used for Control of Aquatic Weeds**

| *Weeds* | *Herbicide* | *Usual dose* | *Method of application* |
|---|---|---|---|
| Ipomoea | 2, 4-D | 2-4 kg/ha | Foliar spraying |
| Vallisneria | 2, 4-D | 10-20 kg/ha | Root zone treatment |
| Ottelia | 2, 4-D | 10-20 kg/ha | Root zone treatment |
| All submerged weeds | Ammonia | 10-15 ppm | Root zone treatment Dispersal in water column |
| Pistia | Paraquat | 0.2 kg/ha | Foliar spraying |
| Salvinia | Paraquat | 0.4 kg/ha | Folliar spraying |

## Biological Control

This method involves the use of biotic agents for aquatic weed control. A number of animals feed on aquatic plants or cause damage resulting in their eradication in fishponds. The grass carp (*Ctenophayngodon idella*) is a very useful fish for biological control of some aquatic weeds. Many species of submerged vegetation *viz.*,

*Hydrilla, Najas, Ceratohyllum* etc. can be controlled by stocking the weed infested waters with grass carps of 20 mm and above size. The surface floating weeds such as *Wolffia, Lemna* and *Azolla* are also consumed by grass carp. The grass carp is also introduced in U.S.A. during 1970 for the control of aquatic submerged weeds on an experimental basis. After 5 years, nearly 20,000 hectares of public lakes of Arkansas was found to be free from most of the submersed weeds. It consumed submersed weeds such as *Hydrilla, Chara, Ceratophyllum,* and *Potamogeton.*

Singh *et al.* (1967) found grass carp of 600 g size to be effective against most submerged weeds under stocking density of 250 to 500/ha. The fish can utilize up to 286 per cent of their initial body weight as observed by them.

The other fishes which are considered useful in controlling some aquatic weeds are *Puntius javanicus, Pulchellus pulchellus, Tilapia mossambica, T.melanopleura,* and *Ophronemus gorami.* However their large-scale use needs further trials. The other animals employed for weed control work are the duck and geese.

In India some snails such as *Lymnaea luteola, Indoplanorbis exustus* and *Pila globosa* are found to be promising for the control of *Salvinia.*

Yamuna turtle, *Kachuga tectum* has been reported to be a consumer of water hyacinth.

## REFERENCES

Hickling, C.F., 1962. *Fish Cultture.* Faber and Faber, London, 295 pp.

Mitra, G.N., 1956. Reclamation of swamps for fishery development in Orissa. *Pregr. Fish. Dev. India,* pp. 54–62.

Philipose, M.T., 1963. Making fishponds weed free. *Indian Livestock,* 1(2): 20–21, 34–36.

Swingle, H.S. and E.V. Smith, 1947. Management of farm fish ponds. *Bull. Ala. Agric. Exp. Sta.,* 264: 34.

*Chapter 8*

# Impact of Exotic Fishes on Indian Fish Fauna

The exotic fishes like *Cyprinus carpio, Ctenopharyngodon idella, Lebistes reticulates, Tina tinca, Hypopthalmichthys molitrix, Tilapia mossambica, Salmo trutta fario, Salmo gairdneri gairdneri, Gambssia affinis, Salvelinus fontinalis, Salmo salar, Onchorynchus nerka, Aristichthys nobilis , Clarias lazera* and *Clarias gariepinus* etc were important and introduced in Indian waters. The transplantation of these fishes into Indian waters has not been a very good experience. The present chapter is collective information on the exotic fishes and their impact on Indian fish fauna.

## *Tilapia mossambica*

Tilapia is native of Africa and the middle East. They have spread mainly through introductions for fish farming and are now found in all tropical and semitropical continents. Tilapia is now being farmed in more than 90 countries. It is cultured in freshwaters of ponds and reservoirs of Java, Malaysia, Thailand, India, Pakistan, Vietnam, South Africa, Uganda, U.A.R., Zambia etc and the backwater ponds of Indonesia, Malaysia, India, Pakistan etc. Some tilapia species can adapt to full strength seawaters also. The nickname 'aquatic chicken' has been used on tilapia, due to its availability to grow quickly with poor quality inputs. In India, the

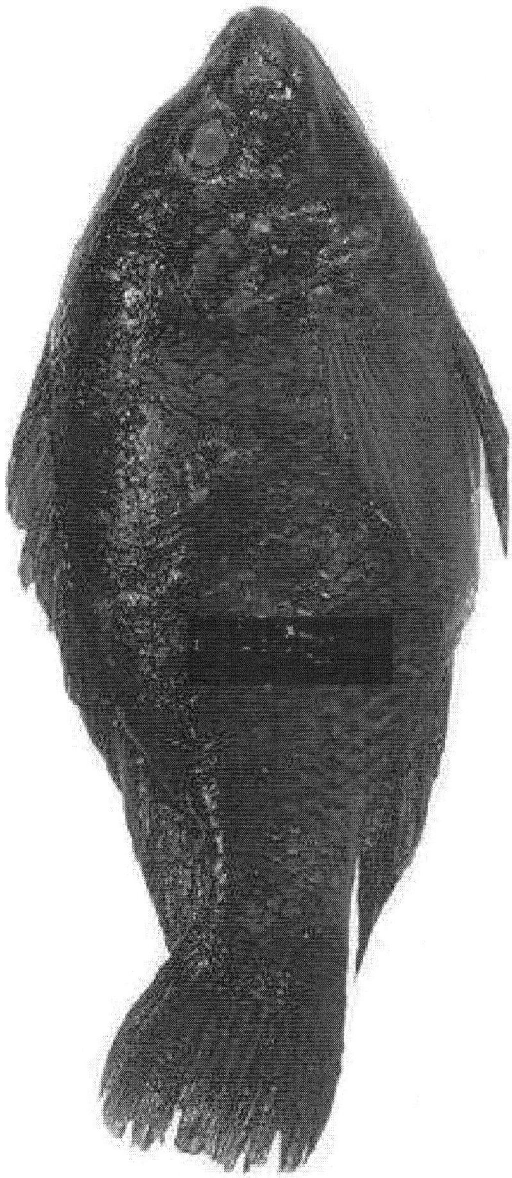

Figure 8.1: *Tilapia mossambica*

first consignment of tilapia was brought by the Central Marine Fisheries Research Institute (CMFRI), on August of, 1952 from Bangkok and the second by the Madras Fisheries Department in the same year from Ceylon.

Tilapia is hardy but display some of the most undesirable characteristics. It bears at a very small size, is difficult to grow to a reasonable market size (100 gm+) with huge differences in growth between the sexes. Due to all these characteristics, tilapias are referred as 'weed fish'. The introductions of tilapia into Indian waters have not been a very happy experience. It has adversely affected the indigenous gene pool.

The presence of tilapia in carp nurseries seriously affected the survival and growth of carp fry since tilapia not only feeds extensively on carp fry but its young compete directly with carp spawn and fry for food. An experiment conducted at Cuttack during 1958-59 to determine the influence of tilapia on the survival and growth of 15 days old fry of rohu and common carp demonstrated that 1) the total fish production in ponds stocked with rohu and common carp fry was more than that of tilapia, rohu and common carp 2) The presence of tilapia population appeared to have brought in a fall in the survival of rohu fry by 17 to 22 per cent and the common carp by 16 per cent 3) In the case of rohu as well as common carp the presence of tilapia seemed to have caused very poor growth, the average individual weight attained being only about half of that in the control ponds, and, 4) In ponds where all the three species were stored together, the presence of rohu and common carp adversely affected production of tilapia to the extent at any 36 per cent as compared to controls, where the presence of tilapia adversely affected production of common carp to the extent of 61 per cent as compared to controls where the presence of tilapia adversely affected production of common carp to the extent of rohu to about 64 per cent.

In Pawai lake of Mumbai the major carps have been badly hit with the accidental introduction of *O.massambicus* with the production sliding down from 33.3 kg/ha to 11.9 kg/ha (Das *et al.*, 1990).

Sreenivasan (1967) found that the growth of rates of *Catla catla*, *Labeo fimbratus* and *Cirrhinus mrigala* were adversely affected by tilapia in Ayyamkulam pond. He also reported that the growth of *Chanos chanos* was restricted to less than 100 gm/year, against the usual 500 gm/year in many water bodies of Tamil Nadu due to co-existence with tilapia. In Kabini reservoir tilapia has adversely affected the indigenous *Cirrhina reba. During* the period from 1980-81 to 1984-85, tilapia has caused decrease of *Cirrhina reba's* share in the catch from 70 per cent to 20 per cent (Murthy *et al.*, 1986)

The accidental introduction of tilapia in lake Jaismand (Rajasthan) has already been reported by fishing chimes. The catch has registered more than two fold increase in month of fishing and the whole catch was only of tilapia. The size of tilapia was found reduced compared to that of early period. It is probably due to the inadequacy of food in reservoir.

Tilapias also form an important fishery of the bheries in West Bengal. Impact of tilapia in the bheries is very severe. The carp fishery that contained up to 93 per cent in 1959 was reduced to nil by 1971, associated with an overall reduction in the yield to the extent of 87.7 per cent.

According to Jhingran (1985) tilapia is unsuitable for culture along with Indian major carps because of the adverse effect it causes on the growth and production of carps and its depredations on carp fry. Although under certain condition, as under monosex culture, it can grow in to a big size, tilapia is not preferred by consumers due to its usually small size. Das *et al.* (1990) also strongly recommended banning of introduction of exotic fish species in our natural water.

## Silver Carp (*Hypopthalmichthys molitrix*)

It naturally occurs in rivers by South and Central China. It is one of the Chinese major carp and has been introduced into Indian waters in 1959 for cultural purpose. It is a surface feeder. Its fry mainly feed on zooplankton, where as, fingerlings and adult feed primary on phytoplankton.

According to Karamchandani and Mishra (1980) silver carp and catla share a common niche and compete with each other for food in a reservoir ecosystem. Percentage composition of phytoplankton in the guts of both the fishes caught during the same time from Kulgarhi reservoir was more or less the same. Zooplankton, the favourite menu of catla, formed 21 per cent of the gut contents of silver carp. The authors concluded that silver carp hampered the growth of catla in the reservoir and advocated caution before its stocking in Indian reservoirs

In Gobindasagar reservoir, stocking of silver carp has since become irrelevant, as the fish has already carved out a nice for itself in the reservoir. In the process, the lurking fear that the exotic fish is deleterious to the population of indigenous catla has been proted beyond any doubt. Ever since silver carp gained a stronghold in the reservoir, catla which constituted an annual fishery of the magnitude of 200 to 300 t has declined considerably.

## Common Carp (*Cyprinus carpio*)

Its original natural distribution was probably restricted to narrow belt of Asia. It has been transplanted into many countries. In India, it was introduced in 1939. There are three varieties of *Cyprinus carpio*.

### Scale Carp (*Cyprinus carpio var. communis*)

Whole body is covered with regularly arranged small scales.

### Mirror Carp (*Cyprinus carpio var. specularis*)

Some parts of the body are covered with irregularly arranged scales.

### Leather Carp (*Cyprinus carpio var. nudus*)

Body is devoid scales.

**Figure 8.2: *Cyprinus carpio***

It is a bottom feeder, omnivorous and feeds on vegetable debris, insects, worms crustaceans and planktonic algae.

Common carp is not suitable fish for stocking in Indian reservoirs, especially the larger ones, for diverse reasons. Being a sluggish fish, its chances of survival in a predator-dominated reservoir are very poor. Due to its slow movement and bottom dwelling habit, they are not frequently caught in a passive fishing gear like gill net. It is no wonder, despite a regular stocking for 13 years, not a single common carp was ever caught from Nagarjunasagar reservoirs of Andhra Pradesh. The stocked fishes failed to survive among the predators. A more important disqualification is its propensities to compete with some important indigenous carps like *Cirrhinus mrigala, C.cirrhasa* and *C. reba* with which common carp shares food niche. The presence of common carp has resulted in the decline of Cirrhinus species in Girna (Maharashtra) and Krishnarajsagar reservoir (Karnataka).

Common carp caused alarming displacement and almost elimination of the prized snow trouts (*Schizothorax spp*) from Kashmir basin by the German phenotype of common carp. The snow trout species biologically handicapped by low fecundity, and ecologically at a disadvantage by the stream breeding behavior which restricts its breeding period of snow-melt period during April-June progressively declined in population density in the face of fierce ecological composition from Common carp which established its pervasive presence all over the lake forming 75 per cent of the catch.

## Thai Magur (*Clarias gariepinus*)

The African sharp tooth catfish, *Clarias gariepinus*, comprising 20 per cent of the total fish catch in Africa, is a remarkable and fascinating catfish species. It is extremely hardy and withstands adverse environmental conditions and habitat instability. It has been translocated to several countries from Africa and is now farmed either in its native or as a hybrid in several countries such as Republic of Korea, Peoples Republic of China, Taiwan, Philippines, Vietnam, Indonesia, Brazil, Poland, Hungary, Belgium, Bangladesh and India. It is widely known for its rapid growth and size. It attains a weight of more than one kg in a year when suitable fed.

Being highly predatory and cannibalistic in nature, the entries of the fish into Indian waters entail a serious effect on indigenous fishes. The fish has however already made an entry into Indian waters through Bangladesh and has gained wide popularity among fish farmers because of its faster growth. Unfortunately, there are highly adverse ecological implications of the proliferation of this fish and many farmers are not aware of these. The culture of the species is not therefore recommended. Studies on larval rearing of this species at Central Institute of Freshwater Aquaculture (CIFA), Bhuvaneshwar indicate that even four days old spawn start chasing and injuring the others causing 2-3 per cent mortality per day as seen in the indoor rearing system.

In West Bengal fish farmers have already suffered a great loss and are ruined by the culture of this species.

*C.gariepinus* is gaining popularity among fish farmers of Punjab as an economically viable alternative species to be cultured for diversification from carp culture system.

## Bighead Carp (*Aristichthys nobilis*)

This fish has been introduced in India without official sanction. Ministry of Agriculture, Government of India, Department of Animal Husbandry and Dairying, Fisheries division, through their letter No.31016/1/96 and FY (3) dated 19th December 1997 had requested all the state governments to destroy the existing stock of magur (*Clarias gariepinus*) and big head carp (*Aristichthys nobilis*).The reason for the ban is that these may cross breed with the endemic species and cause genetic degradation of our fish fauna. National committee on exotic species had not approved the introduction of *A. nobilis* into the country. Indiscriminate stocking of big head carp in ponds had been observed to adversely affect the catla population.

## Grass Carp (*Ctenophanyngodon idella*)

It is a natural of rivers of China. It has been introduced in India in 1959 for culture purpose. It has so far been rather beneficial in culture fisheries. It is believed that this trend will continue as long as the species is not able to establish itself in Indian natural waters because of easy breeding. Grass carp has very limited role-play in the natural waters. However, it has proved to be very effective in the biological control of weeds like *Lemna, Hydrilla, Wolfia, Najas, Chara, Ceratophyllum, Myriophyllum* etc.

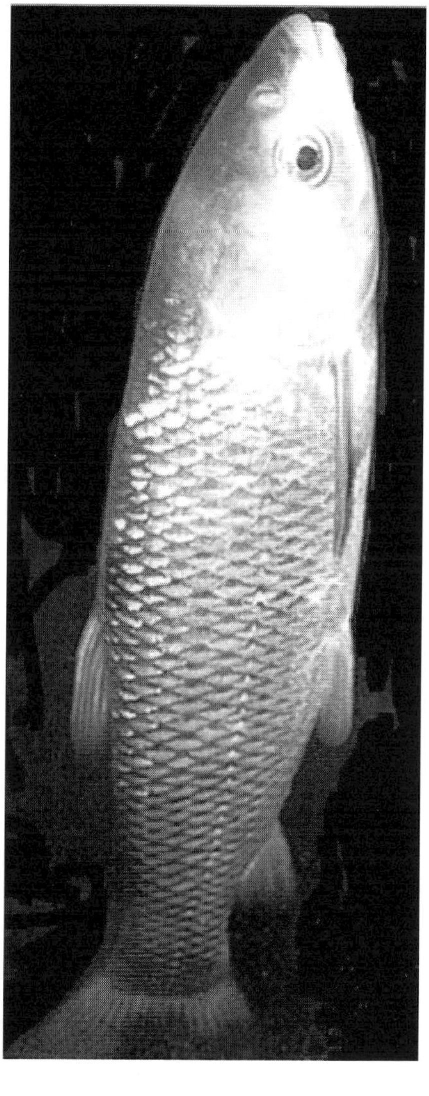

**Figure 8.3:** ***Ctenopharyngodon idella***

## Conclusion

There is an urgent need to educate fish farmers about the consequences of unauthorized introduction of exotic fishes and on the efforts needed for minimizing the risk factors. No exotic fish should be introduced unless it is scientifically recommended and approved. The unauthorized entry of the fish seed of undesirable species should be immediately stopped. It is duty of the extension officers/scientists to educate the farmers about the adverse consequences of culture of exotic fishes.

## REFERENCES

Das, P., G. John and W.S. Lakra, 1990. Exotic fish germplasm in Indian reservoir. p. 45–47. In: Jhingran, Arun G. and V.K. Unnithan (eds.). *Reservoir Fisheries in India.* Proc of the Nat. Workshop on Reservoir fisheries, 3–4 January, Spl. Publ. 3, Asian Fisheries Society, Indian Branch, Mangalore, India.

Dhawan, Asha and Kamaldeep Kaur, 2001. *Clarias gariepinus* in Punjab waters. *Fishing Chimes,* 21(5): 56.

Jhingran, V.G., 1985 *Fish and Fisheries of India.* Hindustan Publishing Corporation (India), Delhi.

Karamchandani, S.J. and D.N. Mishra, 1980. Preliminary observations on the status of silver carp in relation to Catla in the culture fishery of Kulgarhi reservoir. *J. Bombay Nat. Hist. Soc.,* 77: 261–269.

Murthy, H.S., S.S. Gonda and V. Ayyar, 1986. Fisheries of Kabini reservoir in Karnataka. *Fishing Chimes,* 6(7): 36–38.

Sreenivasan, A., 1967. *Tilapia mossambica*: Its ecology and status in Madras state, India. *Madras J. Fish.,* 3: 33–39.

*Chapter 9*

# Estuarine Fisheries

An estuary may be defined as a region where river water mixes with seawater and considerably dilutes it. It is the most variable and dynamic aquatic ecosystem, which is influenced by factors like water movements, tides, mixing of sea and river water, salt load and salinity regime. India has an estimated estuarine waters spread area of 2.6 million hectare with the support of mighty estuaries on east and west coasts.

Diversities in salinity regimes and other physico-chemical conditions due to varying intensities to freshwater discharges in different seasons and tidal fluctuations throughout the year bring about biological alterations in the estuarine ecosystem. The faunistic production of adaptable characteristics can withstand the dynamic stress in estuarine waters and thus only a few species of high commercial value are dominating the catches from open waters and also grown in impounded environments.

## Open Estuarine System

### Hooghly-Matlah Estuarine System

The Hooghly-Matlah estuarine system located in West Bengal between latitude 21°–23°N and longitude 88°–90°E. The Hooghly-Matlah estuarine complex has the largest deltaic region in the world. The total approximate area of this estuarine system with its innumerable tributaries and network of creeks is 2340 sq.km.The mangrove vegetation grown in the deltaic region harbours a potential grazing ground for the estuarine fishes and prawn species. The estuarine sources are highly productive and the total catch from Hooghly-Matlah estuarine system fluctuates within 20,000 – 25,000t/year. Catch assessment from difficult zones of the estuarine complex reflects highest production from the lower zone (zone III) which contributes 70 to 75 per cent to the total annual landings.

Different types of fishing gears are operated in Hooghly-Matlah estuary. Bulk of the total fish landings are contributed by the bag net and the percentage contribution by this type of gear ranges between 70 and 80.

Of special importance in the Hooghly-Matlah estuarine system is the so-called migratory fishing, practiced during winter in the lower zone. Because of the prevailing calm weather in the coastal belt, the fishermen engaged in fishing in upper zone of the estuary migrate to the lower zone for intensive fishing. In the winter, fishing centers like Frazerganj, Jamboodwip, Kalistan etc.the average catch per unit of effort ranges between 29 and 156 kg (1987-88) during November to January, as against 2.6 kg obtained in the upper and lower reaches during the whole year. The important species in bag net catches during winter migratory fishing are *Horpodon nehereus, Settipinna spp, Trichiurs spp* and prawns mainly *Parapenaeopsis spp.*

In early seventies, after the construction of Farakka Barrage an improved level of fishery in the Hooghly estuary was indicated with the release of freshwater into the system through the barrage and the yield in recent years has been on the higher side *i.e.,* around 2000t per year. The yields of purely marine forms (*P.indicus, T.jella, C.ramcarati* and *L.calarifer*) have markedly deceased in the lower zone while fisheries of some species like *P.pama* and *P.paradiseus,* which were in declining trend, have shown a sign of improvement in lower and gradient zone of the estuary.

Wide fluctuation in hilsa catch is an interesting characteristic of Hooghly estuary. However, the landing of the species which remained 1457 tonnes/year during 1966-67 to 1974-75 showed a sharp rise of 2126.2 tonnes/year after the Farakka discharge (1975-78) and has come down again to 1781.3 tonnes/year in recent years (1984-87). The recent upward trend in hilsa landing can be attributed to the increased effort and mechanization of the boats.

## Mahanadi Estuarine System

The Mahanadi estuary lies in Cuttack and Puri districts of Orissa and drains into the Bay of Bengal. The tidal influence extends up to about 42 km from the mouth of the river. As an estuarine complex including the brakishwater lagoons of about 12000 ha, Mahanadi estuarine system covers about 30,000 ha area with an estimated annual production of about 550 tonnes/year.

The fishery of this estuary includes clupeoids, mullets, and prawns. The important fishes present are *Hilsa ilisha, Nematolosa nasus, Sardinella spp.* and *Triisocles spp., Mugil cephalus, M..cunnesius, M.parsia, Lates calcarifer, Sillago spp, Epinephelus spp, Malabaricus spp, Ambassis spp, Leiognathus spp, Lutianus johnii, Datnoides quadrifasciatus* and *Sparus berda.*The 1959-60 survey reported 6900 fishermen actively engaged in fishing activity using 1700 boats and 56700 nets in this estuarine system.

## Estuaries of the Peninsular India

### Godavari Estuarine System

It is the major estuary in peninsular India. It has an area of about 18000 hectares. Goutami is the main source of the estuarine complex in which tidal influence extends only up to 40-50 km upstream from the mouth region. Formation of sand bars in the estuarine mouth restricts the entrance of tidal waters like Mahanadi estuarine system. Detailed studies on zooplankton was made by Prof. Ganapati and his colleagues during 1958-64. About 12 species belonging to 7 genera of mysids, 60 species of copepods apart from 17 genera of medusa and other groups such as isopods, amphipods, siphonophores and ctenophores were encountered. The ecology of the bottom invertebrate fauna with special emphasis on polychaetes, molluscs and echinoderms were studied by several workers. A total of 84 species of polychaetes excluding serpulids, 115 species of gastropods, 57 species of pelecypods and 9 species of echinoids have been identified. The total production from Godavari estuary is estimated to be about 5000 tonnes. Mullets and prawns form the major catch of the system. *M.monoceros, M.dobsoni, M.affinis, M.brevicornis, P.monodon,* and *P.indicus* are the important prawn species available in lower reaches of the estuary.

About 200 species of fishes have been recorded from the estuary, of which 72-80 per cent are nearly euryhaline, 12-20 per cent almost marine and 15 per cent freshwater in origin. Mullets form nearly 2/3$^{rd}$ of the total fish landing and are represented by *M.cephalus, M.speigleri, M.dussumieri, L.troschelii, L.oligolepis,* and *L.seheli.* The other fishes of commercial importance are *Pristiopama hasta, Leiognatus sp, Gerres filamentous, Caranx sp, Sillago sihama, Lates calcarifer* etc.

## Other Estuarine Systems

Other than Godavari estuarine system there are eight small estuaries in peninsular India namely Vasishtha, Vainatheyam, Adyar, Karuveli, Ponniyar, Godilam, Paravan, Vellar, Killai and Coferoom.

The 40 km long the Adyar river meets the Bay of Bengal at Adyar within the southern limits of Madras city. The tidal influence is felt in the river up to a distance of 6.4 km from its mouth. The estuary is connected to the sea only during monsoon and post monsoon months. When freshwater discharge reduces by the end of January, sand bar is formed at the river mouth and the estuary gets totally cut off from the sea. Common phytoplanktons of the Adyar estuary are represented by *Microcystis, Scenedesmus, Oscillatoria, Merismopedia, Asterionella, Coscinodiscus, Campylodiscus* and *Skeletonema.* Zooplanktons are not abundant, except *Brachionus rubens* through out the year. Among important fishes the estuary is having mullets, *Sillago sihama, Tilapia mossambica,* catfishes and prawns.

The Vellar river drains into the Bay of Bengal at Port Novo. The total length of the river is 480 km. The Vellar estuary is located in between latitude 11°-29′N and longitude 79°-46′E.The estuary covers an area of 262 ha. The important fishes of the estuary are *Mugil spp, Leiognathus spp, Sardinella gibbosa, Johnius spp, Polynemus tetradactylus, Engraulis malabaricus,* catfishes and prawns and shrimps. According to Thomas and Raffi (2001) the temperature of estuarine water ranged between 30.5 and 32°C, pH between 8.2 and 8.7 and salinity 13.2 and 82.5 ppm. Maximum value of organic carbon (5.246 mg of C) was recorded in the post monsoon season and the minimum (2.923 mg of C) with a range of 0.69 to 4.83 mg of C during the premonsoon season. The premonsoon recorded the peak values of Cu (110.6 ppm), Fe (48482.1 ppm) and Mn (941.3 ppm); summer recorded increased levels of Cd (129.6 ppm) and Zn (187 ppm). Cr, Ni, Pb and Hg were found in higher levels during monsoon with the values of 460.7

ppm, 579.3 ppm, 194.8 ppm and 535.7 ppm, respectively. The minimum values of Cd (290.0 ppm), Ni (327.6 ppm), Zn (99.7 ppm), Fe (42692.3 ppm) and Mn (871.5 ppm) were recorded in post-monsoon. Monsoon recorded 11.27 ppm of Cu and premonsoon 40.26 ppm, 10.7 ppm and 5.6 ppm of Cr, Pb and Hg as minimum values, respectively.

## Estuaries of the North West Coast

Narmada, Tapti and Mahi are three main estuaries in west coast.

### Narmada Estuarine System

River Narmada is having an estuarine stretch of about 120km which ends in Gulf of Cambey. Gradual decline in tidal ingress is being observed due to development of sand bars at the mouth region of the estuary. However, from ten years records during 1973-82 the average annual production of the estuarine system has been estimated to be about 4000 tonnes in which as a group hilsa contributed (40-45 per cent average 1662.4 tonnes) followed by mullets (15-20 per cent average 686.95 tonnes) and prawns (5-8 per cent average; 286.55 tonnes). Miscellaneous species contributed 30-38 per cent (average 1520 tonnes) of the total landings. The fishing potential estimated in a survey during 1986-87 indicated that 2884 fishermen are actively engaged in fishing activities in the system with 103 boats (nonmechanised) and about 28000 nets. Gill nets are in maximum use (26414 nos.) in the lower zone of the estuary. Hilsa forms the major catch of these gill nets (80-85 per cent). Cast nets are mainly used for prawn fishing. Besides these, the seedling of freshwater giant prawn (*M.rosenbergii*) form an important fishery resource in mixoligohaline and limnetic zones of the system.

On account of a cascade of dams under construction on the Narmada river system with the world bank aid the entire ecosystem and its productivity are bound to be greatly affected. Pre-impoundment surveys have been initiated by the Central Inland Fisheries Research Institute (CIFRI), Barrackpore (West Bengal) in 150 km stretch of the river Narmada in the state of Gujarat.

## Chapter 10

# Fisheries of Lakes

### Fisheries of Pulicat Lake

Pulicat lake is second largest lagoon on east coast spreading in the Nellore district of Andhra Pradesh and the Chingleput district of Tamil Nadu where it is connected with Bay of Bengal by narrow mouth near the Pulicat village. In length it is 60 km and the width varies from 1-18km with total area of 77700 ha but with water spread area of 36900 ha rest being dry and becoming marshy during monsoon. It is connected to sea by a narrow mouth, whose position, depth and width vary with season. Two seasonal rivulets-Rayala vagu and Kalangi river drain into the lake. In the lake, there are several islands the important being Pernadu, Venadu and Irakkam.The average amplitude of the tide at lake mouth is only 25 cm and its influence is felt up to 6 to 10 km from the mouth. The average depth is 1.5 m. During the period 1970-77 total catch varied from 759.6 t (1975) to 1371.4 t (1972) yield ranging from 20.6 to 37.2 kg/ha;of the water spread area. During 1970 prawns accounted for 566.5 t of which the main contributors were *Penaeus indicus* (335.9 t), *M.monoceros* (76.1t) and *M.dobsoni* (54.4t). Mullets accounted for 214.6t the main species being *M.cephalus* (136.2t). Perches amounted to 87.4 t. Clupeids accounted for 125.6t, the main species being *Nematolosa nasus* (70.2t), crabs were represented by 4 *Neptonus pelagicus* (70.2 t) and *Scylla serrata* (48.8 t).

Among the gears used suthuvati (stake net) accounted for about 25 per cent of total catch, badivalai (shore seine) 13 per cent, kondavalai (dragnet) 12 per cent and Kalluvalai (stake net) 11 per cent.

Mullets and prawns are the major fisheries of the lake. Fishes like *Mugil cephalus, M.macrolepis, Sillago sihama, Gerres oyena, Nematalosa nasus* and *Chanos chanos*; prawns like *Penaeus semisulcatus, P.monodon, P.indicus, Metapenaeus monoceros* and *M.dobsoni*; crabs like *Scylla serrata* and *Portunus pelagicus* and oyster like *Crassostrea madrasensis* are also found in the lake.

## Fisheries of Chilka Lake

The Chilka lake located along the east coast of India between the latitudes 19028'-19024'N and longitudes 85067' and 85035' E is a pear shaped brakishwater lagoon in between the district of Puri, Kaurda and Ganjam has an average water spread area of 1050 sq.kms.It covers an area of 1165 sq.kms. in monsoons and 906 sq.kms. in summer.

The lake is fed with main two tributaries of river Mahanadi called river Daya and river Bhargavi at the northern side and connects with the Bay of Bengal near Arakhakuda village through a zig zag long outer channel of 35 kms long. Similarly the lake has connected with the Bay of Bengal at the southern end near Purunabandha village through 16 kms long canal called as 'palur canal'. As a result of this, the entire lake becomes a freshwater during monsoons. This peculiar phenomenon change provides a congenial features for survival of freshwater flora and fauna up to December and afterwards the salinity gradually increases touching maximum in July.

## Hydrography and Physico-chemical Characteristics of Lake

The depth of the lake is quite uneven, the northern part being shallowest (47-134cm) and the southern and some parts of the central sectors are comparatively deep (102-300 cm). The water level remains minimum during April to June reaching highest during the monsoon period.

The transparency of water showed comparatively high values (70-140cm) all along the western bank low values (5-25 cm) were observed on the eastern bank of the northern and central sectors during monsoon period.

Annual rainfall at Chilka is recorded at 114-117 cm. The water temperature of the lake varies from 17.5°C in January to 32°C in July.

The salinity shows sharp annual cyclic changes in the range of 0.13 to 36.02 per cent. The salinity starts rising from November and reaches high values in April-June period in both the central and northern sectors. In the southern sector, however, the fluctuation of salinity is comparatively less pronounced.

Dissolved oxygen ranged between 2.6 to 15.6 ppm and total alkalinity from 25.8 to157 ppm.pH was in the range of 6.8 to 9.7. Higher values of phosphates, nitrates, silica and iron occurred in monsoon months while that of total alkalinity in summer months.

## Flora and Fauna

Lake is heavily infested by the aquatic weeds such as *Phanerogams potamogeton pectinatus, Najas fareolata, Halophila ovata, Ruppia martitima* and *Halopila baccavii*.The algae in the lake include the marine species *Gracilaria edulis* and *G.verrucosa* and the fresh species *Hydrilla verticillata* and *Vallisneria spiralis* in the southern and central sectors are of economic importance.

The faunal diversity comprises few species of one sea anemones, thirty-seven species of nematodes, one species of polychaetes, seven species of stomotopods, twenty-eight species of crabs, sixty species of mollusca, sixty-nine species of fishes, 150 species of birds, and one species of dolphin.

The lagoon had been facing problems like siltation, shrinkage of area, chocking of the inlet channel as well as shifting of the mouth connecting the sea, decrease in salinity, proliferation of invasive freshwater species, decrease in fish productivity and an overall loss of biodiversity.Due to changes in the ecological characters of the lagoon it was placed in the Motreux record in the year 1993.While comparing the data contained in the pioneering studies on Chilka done during the period 1915 to 1924, with the data generated by way of extensive survey carried out by Zoological Survey of India during 1985-87, it was opined that lagoon ecosystem was tending towards freshwater ecosystem and warranted urgent restoration measures.

**Table10.1: Fish Species in Chilka Lake**

| Sl.No. | Species | Family | Habitat |
|---|---|---|---|
| 1. | *Ailia coila* | Schilbeidae | Pelagic |
| 2. | *Ambassis commersonii* | Ambassidae | Benthopelagic |
| 3. | *Anabus testudineus* | Anabantidae | Demersal |
| 4. | *Aplocheilus panchax* | Aplocheilidae | Benthopelagic |
| 5. | *Bagarius bagarius* | Sisoridae | Benthopelagic |
| 6. | *Caranx sexfasciatus* | Carangidae | Reef associated |
| 7. | *Catla catla* | Cyprinidae | Benthopelagic |
| 8. | *Chanda nama* | Ambassidae | Benthopelagic |
| 9. | *Channa punctata* | Channidae | Benthopelagic |
| 10. | *Channa striata* | Channidae | Benthopelagic |
| 11. | *Chanos chanos* | Chanidae | Benthopelagic |
| 12. | *Chela laubuca* | Cyprinidae | Pelagic |
| 13. | *Clarias batrachus* | Clariidae | Demersal |
| 14. | *Crossocheilus latius* | Cyprinidae | Benthopelagic |
| 15. | *Cynoglossus lingua* | Cynoglossidae | Demersal |
| 16. | *Echeneis naucrates* | Echeneidae | Reef associated |
| 17. | *Elops machnata* | Elopidae | Pelagic |
| 18. | *Esomus dancricus* | Cyprinidae | Benthopelagic |
| 19. | *Etroplus suratensis* | Cichlidae | Benthopelagic |
| 20. | *Gazza minuta* | Leiognathidae | Demersal |
| 21. | *Gerres filamentosus* | Gerreidae | Demersal |
| 22. | *Gerres setifer* | Gerreidae | Demersal |
| 23. | *Glossogobius giuris* | Gobiidae | Demersal |
| 24. | *Gudusia chapra* | clupeidae | Pelagic |
| 25. | *Heteropnestus fossilis* | Heteropneustidae | Demersal |
| 26. | *Hilsa kelee* | Clupeidae | Pelagic |
| 27. | *Labeo ariza* | Cyprinidae | Benthopelagic |
| 28. | *Labeo calbasu* | Cyprinidae | Demersal |
| 29. | *Labeo rohita* | Cyprinidae | Benthopelagic |
| 30. | *Lates calcarifer* | Centropomidae | Demersal |
| 31. | *Liza macolepis* | Mugilidae | Demersal |
| 32. | *Liza parsia* | Mugilidae | Demersal |

*Contd...*

**Table10.1–Contd...**

| Sl.No. | Species | Family | Habitat |
|---|---|---|---|
| 33. | *Liza tade* | Mugilidae | Demersal |
| 34. | *Macrognathus aral* | Mastacembelidae | Benthopelagic |
| 35. | *Mastacembelus armatus* | Mastacembelidae | Demersal |
| 36. | *Megalops cyprinoides* | Megalopidae | Pelagic |
| 37. | *Mugil cephalus* | Mugilidae | Benthopelagic |
| 38. | *Mystus cavasius* | Bagridae | Demersal |
| 39. | *Mystus gulio* | Bagridae | Demersal |
| 40. | *Mystus vittatus* | Bagridae | Demersal |
| 41. | *Nematalosa nasus* | Clupeidae | Pelagic |
| 42. | *Notopterus notopterus* | Notopteridae | Demersal |
| 43. | *Ompak bimaculatus* | Siluridae | Demersal |
| 44. | *Pangasius pangasuis* | Pangasiidae | Benthopelagic |
| 45. | *Plotosus canius* | Plotosidae | Demersal |
| 46. | *Plotosus lineatus* | Plotosidae | Reef associated |
| 47. | *Pristis pectinata* | Pristidae | Demersal |
| 48. | *Pterois radiata* | Scorpaenidae | Reef associated |
| 49. | *Puntius sophore* | Cyprinidae | Benthopelagic |
| 50. | *Puntius ticto* | Cyprinidae | Benthopelagic |
| 51. | *Puntius vittatus* | Cyprinidae | Benthopelagic |
| 52. | *Rasbora daniconius* | Cyprinidae | Benthopelagic |
| 53. | *Rasbora rasbora* | Cyprinidae | Benthopelagic |
| 54. | *Rhinomugil corsula* | Mugilidae | Pelagic |
| 55. | *Salmostoma bacaila* | Cyprinidae | Benthopelagic |
| 56. | *Sardinella melannura* | Clupeidae | Pelagic |
| 57. | *Scoliodon laticaudus* | Carcharhinidae | Demersal |
| 58. | *Silonia silondia* | Schilbeidae | Demersal |
| 59. | *Sphyraena putnamae* | Sphyraenidae | Reef associated |
| 60. | *Tenualosa ilisha* | Clupeidae | Pelagic |
| 61. | *Terapon jarbua* | Tetrapontidae | Demersal |
| 62. | *Tetradon cutcutia* | Tetradontidae | Demersal |
| 63. | *Wallago attu* | Siluridae | Demersal |
| 64. | *Xenentodon cancila* | Belonidae | Pelagic |

Considering the sensitive nature of the lagoon ecosystem, extensive hydrological studies were carried out by commissioning the services of experts from reputed institutes of the country to identify the best options to restore its ecosystem. Based on the findings of the extensive studies, several ameliorating measures were recommended by these institutes and the Chilka Development Authority (CDA) take steps to execute them to restore its ecosystem.

A two day international workshop on 'Eco-restoration of Chilka lagoon' was organized by CDA at Bhubaneswar.The workshop was held from 19 to 20 January, 2002.The prime objective of the workshop was to conserve and restore the lagoon's ecosystem and at the same time facilitate its sustainable use for promoting economic well being of the local community dependent upon the lagoon for its livelihood and to generate clear recommendations regarding actionable strategies in all needed areas for conservation. Some important recommendations made in the workshop were:

1. In the context of the increase in tidal flow and the salinity observed in the main lake, the variation has to be monitored for the coming years and the real impact can be assessed there upon.
2. The implementation of efficient river management system may be initiated.
3. The study on the littoral drift in ocean side and by passing around the new mouth can be undertaken.
4. Remote sensing application can be effectively used to enhance the strength of the field data collection.
5. Detailed mapping and survey of biodiversity and continuous monitoring may be undertaken.
6. Measures to protect catchment and to prevent sedimentations may be taken.
7. Regular monitoring of floral and faunal diversity necessary may be undertaken.
8. All Dolphin carcasses be preserved and studied.
9. Khanda fishing in Magarmukh and Nalaban be stopped including capture of juveniles. Palur canal be deepened up to the melting point with the open sea.
10. Prawn culture and conversion to agriculture be stopped.

11. Sustainable fish yield and balancing between ecological and economic aspects should be aimed at.
12. Gender issues to be addressed for sustainable fisheries development.

## Fisheries of Kolleru Lake

Kolleru lake is the second largest freshwater ecosystem in peninsular India. It is located between 16°32′ latitudes of 16°47′N and longitudes of 81°2′ and 81°4′E in the state of Andhra Pradesh.The lake was formed as a natural depression between river Godavari on the north and river Krishna on the south near their tail-end region, before they open into Bay of Bengal. Several irrigation canals originating from the dams on river Godavari at Dhowlaiswaram and river Krishna also open into the lake. Lake also receives water through a number of large and medium sized, seasonal rivulets and streams that drain the surrounding upland areas. The lake is connected to Bay of Bengal through a narrow natural channel upputeru, about 62 km long.

The water spread area of 901.3 sq.km. at maximum and 135 sq.km at 1m (M.S.L.).The total catchment area of the lake was estimated to be about 4700 sq.km.Depending upon the seasonal inflow the depth of lake varies between 1m to 4m. During summer the marginal shallow areas of the lake become completely dry.

### Climate

The lake area receives annual average rainfall of 99.95 cm. The relative humidity varies between 64 per cent in June and 83 per cent in February. The lake experiences brakishwater conditions in the months when the inflow of freshwater into the lake is low. The saline water reaches up to the middle regions of the lake, while in the northern parts freshwater conditions exist. On the onset of southwest monsoon in the mid seasonal drains and streams.

### Biodiversity

Kolleru lake supports very rich aquatic flora and fauna. The lake is a eutrophic body with huge quantities of submerged, floating and emergent aquatic vegetation. The lakebed is completely covered with rooted macrophytes. The common species are *Ipomoea aquatica,*

*Salvinia auriculata, Eichhornia crussipes* and *Pistia stratiotes*. Among zooplanktons rotifers, chonchostracans, cladocerans, copepods, cyclopoides, harpacticoides and ostracods are dominant. 11 species of birds have been recorded from lake area.

Near about 105 species of fish and 7 species of commercially important prawns were recorded so far from the lake, out of which 19 species of fish were encountered in brackish waters in the tidal zone. The lake supports a rich fishery of air breathing fishes. Traditionally live fish captured in the lake area exported to markets in the north-eastern India. The Indian climbing perch (*Anabus sp*) dominate the catch. The commercially important catfish include *Clarias batrachus, Heteropnesutes fossilis* and *Wallago attu*.The murrels such as *Channa striatus* and *Channa punctatus* supports a good fishery. Carp landings are represented by *Catla catla, Labeo rohita, Cirrhinus mrigala*, and *Puntius sarana sarana*. Other important fish species are *Mugil cephalus, Etroplus suratensis, Etroplus maculates, Mastacembelus armatus armatus, M.pancalus, Macrognathus aculeatus* and *Anguilla bengalensis*. The commercially important prawns are *Penaeus monodon, Penaeus indicus, Metapenaeus monoceros, M.dobsonii, Macrobrachium rosenbergii, M.rude*, and *M.macolmsoni*.

## Changes in the Lake Habitat

The marginal areas of the lake have been converted into fishponds. World Banks extended financial assistance to farmers for the construction and management of fishponds. In the last decade nearly 24000 hectares area have been converted into the fishponds. This has resulted in the degradation of natural habitat in the lake proper over the years affecting the fishery. The area of the lake has diminished and the breeding areas of the many species are affected.

Dutt and Murthy (1976) studied the fisheries of Kolleru lake for 1970-72, when the ponds were not constructed. During this period the total fish landings were estimated to be around 720 tonnes. Of these fish *Anabus oligolepis* and *A.testudineus* dominated the landings forming nearly 28 per cent. Carps contributed to 18 per cent and murrels to about 16 per cent. Rao *et al.* (1987) based on a study of the landings to be around 1800 to 2800 tones. The total landings during the period between 1976 and 1987 increased three fold. It is mainly due to an increase in the fishing activity in the lake.

**Table 10.2: Major Landings of Fish (in per cent of the total fish landings) per cent in the Catches from the Lolleru Lake**

| *Major Groups* | *Before Construction of Fish Ponds (Dutt and Murthy 1976)* | *After Construction of Fish Ponds (Rao et al., 1987)* |
|---|---|---|
| Perches | 28.37 | 35.39 |
| Catfish | 13.19 | 32.37 |
| Murrels | 16.61 | 17.01 |
| Carps | 18.75 | 10.73 |
| Mullets | — | 1.38 |
| Spiny eel | 22.80 | 0.99 |
| Eels | 1.00 | 0.14 |

A comparative study of the major fish groups on the fisheries of Kolleru lake between the periods before and after the changes in ecology suggests perceptible changes in the fisheries of Kolleru lake. Catfish landings when compared to earlier period also increased. The landings of the climbing perches and murrels remained stable, while that of carps declined remarkably. This could be attributable to the decrease in the lake area, where the juveniles are distributed and also the increased capture of juveniles of carps for stocking in the near by fish ponds. Normally the juveniles got recruited from the streams and irrigational channels opening into the lake. Increased exploitation of juveniles could be one of the reasons for the decline in fishery.

## Fisheries of Dal Lake

The Kashmir valley has a number of freshwater lakes notably Dal lake, Wular Lake, Haigam Rakh, Mirgund and Hokarsar.These lakes are unique in their flora and fauna, their beauty and their biological production. The Dal lake is situated within the Srinagar city lying with a latitude of 34.7°N and longitude of 74.52°E has total open water area of 124 sq.km.

### Plankton Diversity

The phytoplanktons are represented by chlorophycceae, bacillarophyceae and myxophyceae. About 150 phytoplankton genera have been identified. The maximum population of phytoplankton has been recorded in May to June and minimum in December to January. The secondary peak occurs in the month of September.

Copepoda, rotifera and cladocera are the principal components of the zooplankton. Copepoda forms the largest group, next being rotifera and third cladocera, all showing maximum population in spring and autumn and poor population in summer and winter. Ostracoda, having 2 genera, has been found absent in the late spring, late summer and early autumn.

## Fish Fauna

Thirty seven species of fish have been recorded and important species inhabiting the Dal lake are *Schizothorax esocinus, Schizothorax niger, Schizothorax curuifrons, Labeo dero, Crossocheilus sp, Botia birdi, Nemacheilus kashmirensis, Gambussia afffinis, Glyptothorax, Glyptosternum* etc.

Besides the above mentioned fishes, the exotic fishes like scale carp (*Cyprinus carpio communis*) and mirror carp (*Cyprinus carpio specularis*) constitute the major catch of commercial fishery of Dal lake. Common carp introduced in the year 1957, constitutes about 70 per cent of the total catches where as *Schizothorax* contributes for about 10 per cent and other fishes including catfishes form 20 per cent of total landings.

## Gears and Crafts

Cast net is the main fishing gear used in the lake. It is of 6 pieces with 4.8 meters in diameter having mesh size 1 to 2 cm, bar to bar with iron or lead sinkers along the peripheral cord weighing about 5 kg per net. One fisherman operates the net from boat while another fisherman rows the boat. Besides this, scoop net and hook and line are most commonly used by the fishermen. Narsoo, a multiple head spear having 5 to 7 heads, is also used. The nets are fabricated locally by the fishermen, mostly of nylon twine, but cotton twine is also used.

The fishing craft consists of open wooden boat of about 20 feet length and 4 feet center width, having capacity of carrying 2 to 3 persons, besides fish catch of about 50 kg. The boats are mostly constructed from devdar wood.

## Pollution

The factors responsible for pollution in the Dal lake are sewage disposal, supply of nutrients from rice fields, catchment areas, urban surrounding and numerous house boats. The mortality of endemic fish viz., *Schizothorax* and *Botia* is suspected to be due to a large extent to the accumulation of residues of pesticides used for the control of pests of fruits, orchards and floating vegetable gardens in and around the Dal lake and it has been stressed that pesticides and insecticides should be carefully and judiciously used.

The cause of deterioration of lake is due to the excessive inflow of nutrients, overgrazing and encroachment of lake area and wastes from surroundings and houseboats. Removal of excess weeds from the lake is essential. It can be done by using grass cutters or by chemicals, which would kill the weeds but not the fish. The weeds can also be controlled by introducing the Grass carp (*Ctenopharyngodon idella*).It is a hardy, fast growing fish. It may not compete with the existing fish fauna of the lake. Grass carp effectively controls *Azolla, Salvinia, Hydrilla, Najas, Ceratophyllum, Vallisneria, Utricularia, Myriophyllum, Potamogeton* etc which form the common weeds in the Dal lake. These weeds not only use up the plant nutrients in the water but also shade up sunlight, causing less production of phytoplankton. This in turn causes less zooplankton production, as the zooplankters are consumers of phytoplankton. Consequently this leads to starvation or low diet of the fish larvae, fry and fingerlings, 50 to 80 per cent of these endemic fish juveniles dying out in the lake. Then again, pollution kills off large numbers of these juveniles. Pollution in Dal lake can be prevented by special arrangements to pipe off the sewage and excreta, which falls into the lake through the flush systems of over 2000 houseboats all over the lake. This sewage may be collected and converted easily into fertilizers.

## Fisheries of Loktak Lake

It is a shallow and somewhat acidic freshwater lake situated at Lat.24.50°N and Long.93.80°E in the north-east of India in the state of Manipur.The water spread area of the lake is about 390 sq.km. during the flood time. Although relatively shallow, it has a few points where the water depth goes up to 25 feet. The lake is fed by a number of small streams such as Nembal, Thangia, Orok Ning Thou, Khorg turel, phubala turtle etc. It is known for its unusual infestation with weeds, both dead and live, and also heavy siltation. This infestation piled up over years into various floating shapes, oblong, circular, rectangular etc., with a depth of 5 to 6 feet. These are known as 'phums', which cover about half of the surface of the lake. Some of the floating macrophytes are *Lemna, Utricularia, Trapa, Azolla, Pistia, Nymphoides, Wolffia, Salvinia* and *Eichhornia*.Some of the submerged weeds are *Hydrilla, Vallisneria, Potamogeton, Najas, Myriophyllun* and *Ceratophyllum*.The common rooted floating weeds are *Nymphaea, Ipomoea, and Nelumbo*; some shallow water weeds are *Colocasia, Marselia* and *Limnophila*.The margin inhabiting species are *Cyperus, Panicum, Luduirgia* etc.

The main forms among phytoplankton are *Euglena, Cyclotella* and *Dinobryon*.The dominant forms among zooplankton are *Keratella, Diffugia, Brachionus, Diaptomus, Polyarthra, Ceriodaphnia, Bosmina, Cyclops*, and *Moina.*

The common fishes recorded in the lake are *Cirrhinus mrigala, Puntius spp, Catla catla, Labeo calbasu, L.gonius, L.rohita, Cyprinus carpio, Hypopthalmicthys molitrix, Ctenopharyngodon idella, Tilapia spp, Wallago attu, Clarias batrachus, Heteropnestus fossilis, Channa spp, Anabus testudineus, Ambassis sp, Glossogobius giuris, Notopterus notopterus, Osteobrama cotio* and *Esomus dendricus.*

In order to utilize the potential of the lake for beneficial purpose, the state government has set up Loktak Development Authority. While the main purpose of the Authority was to set up a hydro-electirc project, the Authority is also charged with the responsibility of development of fisheries of the lake. As a part of this assignment, it has taken up the task of clearing the lake from 'phums' and scattered weeds. Further, to facilitate stocking, the authority has set up two major carp hatcheries and is understood to have been advised by the fisheries development to raise grass carp seed at the hatcheries for stocking the lake.

A survey of the fishing gears used in the lake revealed that hook and line, pole and line, dip net and ripping types of fishing are practiced in the lake.

## REFERENCES

Badapanda, H.S., 2002. Planned eco-restoration of Chilka lagoon. *Fishing Chimes*, 22(4): 20–23.

Dixitulu, J.V.H., 2002. On Loktak lake fishery development (editorial). *Fishing Chimes*, 22(4): 4–5.

Dutt, S. and V.S.R. Murthy, 1976. On the fish and fisheries of the lake Kolleru, Andhra Pradesh. *Mem. Soc. Zool.*, Guntur, 1: 17–21.

Jayadev, W. and W. Vishwanath, 2002. Fishing practices at water inlet area of Loktak hydel project. *Fishing Chimes*, 22(4): 57–58.

Jaya Raju, P.B., 2000. Ecological wonders of India: Ecological degradation due to intensive fish culture practices in lake Kolleru, Andhra Pradesh, India. *Hydrosphere*, 6: 40–46.

Rao, K.V., R.S.Panwar, M. Rama Krishnaiah, J.B.Rao, T.S. Rama Raju, K.S. Rao and G.Ch. Rao, 1991. Studies on ecology and fisheries of Kolleru lake ecosystem and development of suitable management measures for obtaining sustained fish production. Annual Report, Central Inland Fisheries Research Institute, Barrackpore, p. 26–30.

*Chapter 11*

# Reservoir Fishery of India

The hydraulic structure across a river to store water on its upstream side, which is an impervious or fairly impervious barrier across a natural stream so that a reservoir is formed. This water is then utilized as and when it is needed. Due to construction of reservoir, water level in that river at its upstream side is very much increased, and a large area may be submerged depending upon the water spread of the reservoir so formed. Due to increased demands for reliable supplies of electric power, irrigation, and drinking water, the number of new hydropower reservoirs is increasing dramatically, especially in Asia.

It has been estimated that annual inland fish production in Asia is 5.5 million tons, comprising 57 per cent of the world's inland fish production (De Silva, 1988, 1992). However fish yields from Asian reservoirs comprises just 0.5 million tons of this 5.5 million tons (De Silva, 1988). De Silva (1988) estimated fish production from Asian reservoirs at only 20 kg/ha /year with a wide variability in production that was not always related to the size of the reservoir. Costa–Pierce and Soemarwoto (1987) estimated an average percentage increase in reservoir area in 15 Asian nations from 1987 to 2000 would be 511 per cent ranging from 50 per cent (Singapore) to 9,900 per cent (Laos). By 2000 it is predicted that the collective water surface in reservoirs (20.3 million ha) will exceed the surface area of Asia's natural waters (18.5 million ha). Clearly if the huge expanse of under utilized water areas locked behind Asia's dams could be utilized for increased fish production, thousands of tons of new aquatic protein could enter the Asian markets (Costa-Pierce, 1998). Production of new aquatic protein is especially urgent in Asia, a region where fish is the most important source of animal protein. In addition, there is a need to create thousands of new rural jobs due to population growth is evident and to final innovative ways to stem the rapid rural to urban migration. It is argued that expansion of aquatic food production in Asian reservoirs could assist in mitigating Asia's growing food and population crises (Brown, 1997).

Exposed to the warm tropical climate, these water bodies in our country are extremely productive and they harbour an enviable spectrum of fish genetic resources. Optimum utilization of these resources can lead to a manifold increase in inland fish production, earning the country a place among the top inland fish producing nations of the world. As a result of the construction of dams under various multipurpose river valley projects, a number of fresh water impoundments have come into existence. These impoundments are called 'reservoirs' or 'dam-lakes' which represent the complex environmental and biological systems. Within a reservoir there is a great diversity of interactive components ranging widely in morphometrical, operational and biological characteristics.

Reservoirs are generally classified as small (<1000 ha), medium (1000-5000 ha) and large (>5000 ha), especially in the records of the Government of India (Srivastava *et al.*, 1985). Jhingran and Sugunan (1990) classified the reservoirs as large (>1000 ha), medium (500-1000 ha) and small (< 500).

**Table 11.1: Distribution of Small, Medium and Large Reservoirs in India**

| | *Small Reservoirs* | *Medium Reservoirs* | *Large Reservoirs* | *Total* |
|---|---|---|---|---|
| Number | 19134 | 180 | 56 | 19370 |
| Area (ha) | 1485557 | 527541 | 1140268 | 3153366 |

Data pulled from Sugunan (1995): FAO Report.

The total water spread area of Indian reservoirs is about 3 million hectares, which forms about 50 per cent of the total reservoir area in

South East Asia (Shetty 1990). Indian Institute of Management, Ahmedabad (Srivastava *et al.,* 1985) listed 975 reservoirs in India, in size range of 1000 ha to 10,000 ha covering an area of 1.7 million hectares. According to Sugunan (1995), the present total area under the reservoirs in our country is 3.1 million hectares (Table 11.1).

According to Dixitulu (1999) there are 19, 370 reservoirs in the country with an extent of 3, 153, 366 ha (consisting of 19, 134 nos of small ones with an extent of 1, 485, 557 ha; 180 nos of medium ones with 575, 541 ha and 56 nos of large ones with an extent of 1,140,268 ha) Kaur and Dhawan (1997) mentions total number of 550 reservoirs with a total water spread area of about 2 million hectares. Sugunan (1995) estimated the present annual yield per ha from small reservoirs (less than 1000 ha each) at an average of 49.90 kg, from medium reservoirs at 12.30 kg and from large reservoirs at 11.43 kg. He gives the pooled production for all the three categories of reservoirs as 20.13 kg/ha. Based on these estimates the present total annual production from Indian reservoirs has been computed at 93, 650 t, against the potential of 245,134 t (from 3,153,366 ha); or in other words, the average per hectare potential production has been visualized is around 80 kg/ha/yr (Dixitulu, 1999). He further mentioned that going by this formulation, 245,134 t (from 3,153,366 ha). In other words, the average per ha potential production the export has been visualized is around 80 kg. Going by this formulation 245,134 t should generate gross returns of Rs. 735 crores at an average realization of Rs. 30 per kg as pooled value, realized out of domestic sales and exports through centralized storage plants and the connected cold chain and domestic marketing system. The present Indian reservoir fish production is frustratingly low; the various estimates of national average ranging from 6 to14 kg/ha/yr (Jhingran, 1985). Indian reservoirs, being in the tropics, have high primary productivity and have the capacity to produce more fish than their present low Indian average of 20-25 kg/ha/yr in large reservoirs and 49.5 kg/ha/yr in minor reservoirs (Piska 1999). It is also estimated that the reservoir fisheries development has the potential to generate additional national income to the tune of Rs. 100 crores per year and can provide employment to millions of fishermen and workers in the ancillary industries (Srivastava and Reddy 1983). Dwivedi (1980) states that if the optimum, productivity level is achieved from present water area of reservoirs alone can provide more than 50,000 tonnes of fish and can provide employment to nearly 72,000 fishermen. Khan *et al.* (1991) mentions 3 million hectare reservoir area in our country offer tremendous scope for augmenting fish production; and reservoir fishery is also important from sociological point of view with respect to providing employment to about 2 million people.

Unfortunately, majority of reservoirs are not being scientifically managed for fisheries. Only a handful of them have so far been harnessed along scientific lines, while the others are either half-heartedly managed or not managed at all. As compared to several developed countries, the per hectare fish production in Indian reservoirs is very poor, being only about 20-25 kg per year as against 88 kg in the USSR (Jhingran, 1985), 100 kg in Sri Lanka and 64.5 kg in large reservoirs of Thailand (Bhukaswan, 1977). Dixitulu (1999) mentioned that in case of reservoir fisheries, however, the developmental process has been tardy, remaining by and large, at the level as at the time the British withdrew, with the exception of a few reservoirs. It is to the credit of the British that, at quite a few reservoirs, they set up farms and also undertake stocking of the reservoirs. The process no doubt continued after independence, with the production being conspicuous in a very few reservoirs, below average in some, and low to very low in others. One of the main factors responsible for low fish yield from Indian reservoirs is the unscientific management practices, which stem from the inadequate knowledge of the ecology and production functions of this biotope. Apart from the urgent need to take up all old reservoirs for proper appraisal of their present status and for taking up rational management and conservation measures, it is highly imperative to take up fisheries development work in new reservoirs right from the initial stage itself. Lot of progress has been made in several developed countries in deep water fishing in reservoirs and management of their fisheries. The neglect of this vast potentially rich fishery resource is all the more applling, in view of inadequate fish production in the country even to meet the protein demands of our exploding populations, let alone the export demands.

## Reservoir Fisheries in Different States

### Tamil Nadu

Except in some states like Tamil Nadu it is difficult to get data on even total production of a state. In southern region of country, in

Tamil Nadu the total annual fish production potential for reservoirs is about 2,000 tonnes, which indicates per unit area production of 40 kg (Sanskarasubbaiyan and Menon 1984). However, Sreenivasan (1998) mentions 52,000 hectare of reservoir area. Sugunan (1995) estimated 52 small, 8 medium and 2 large reservoirs with 48.50, 13.74 and 12.66 kg/ha/yr fish yield respectively. CIFRI (1998) surveyed Amaravathy, Palar–pornthalar, Uppar, Pillor, Gunderipallam and Varattupallam reservoirs of Tamil Nadu and revealed that the fish yield in Varattupallam was highest and lowest in Amaravathy. Taking into consideration the area of reservoirs and stocking density, it was concluded that the fishery management in Palar–Poranthalar is best followed by Amaravathy and uppar, while it was poor in Pillar. The actual fish yields obtained from the reservoirs are less compared to the production potentials estimated through primary production studies. This gives scope for enhancement of fish production from these reservoirs through judicious stocking and exploitation.

## Kerala

Being essentially a marine fisheries state, inland fisheries till recently did not receive much attention. There are 29,660 ha of reservoirs with a fish yield less than 15 kg/ha; mostly Tilapia (Sreenivasan 1998). According to Sugunan (1995), state is having 7 small and 2 medium reservoirs with fish production of 53.50 and 4.80 kg/ha respectively. Further Sugunan also mentions, there are total 30 reservoirs with water-spread area of 29,635 hectare. Mohan(2001) highlighted the economic and social conditions of fishermen of Malampuzha reservoir and concluded that 79.17 per cent of respondents were literate.

## Karnataka

Karnataka is an important maritime state, the marine fisheries in state is more dominant than inland fisheries, where a major finance has been expended for the development of marine fisheries. State is having 60 reservoirs with an area of 2.20 lakh hectares (Annual Report, Department of Fisheries 1994-95). However, IIMA (1985) mentions presence of 28 reservoirs with total water spread area of 1,51,624 hectares, the average fish production from which is about 23-35 kg/ha/yr. Sugunan (1995) mentions the total area under reservoir in the state is 4,37,291 hectares comprising 4,679 reservoirs of different categories. Sukumaram (2001) studied status of plankton from Ghataprabha, Malaprabha, Manchenebele, Nugu, Hemavathy, Kabini, Linganamakki, Harangi, V.V.Sagar and Narayanpur reservoirs of the state and concluded that all the reservoirs barring Ghataprabha are free from stress.

## Andhra Pradesh

It has 2.143 lakh hectares of reservoirs of all sizes (Sreenivasan 1998). According to Sugunan (1995), the state is having 37 small, 29 medium and 3 large reservoirs having 188, 22 and 16.80 kg/ha fish yield respectively. IIMA (1985) estimates the average reservoir fish production of 37.43 kg. CIFRI (1998) also conducted the survey of nine reservoirs of Andhra Pradesh and found that most of the reservoirs have the necessary infrastructure for raising the seed and exploitation of the fishery and recommended that the cooperative societies need to be activated and given the responsibility of management. There is an immense scope to introduce cage culture systems in most of the reservoirs. This will enhance the yield from reservoirs and also provide additional employment.

## Rajasthan

It is with 78 small, 17 medium and 2 large reservoirs with a fish yield about 46.43, 24.47, and 5.30 kg/ha/yr respectively (Sugunan 1995). According to IIMA (1985), the total number of reservoirs in the state is 665.

## Uttar Pradesh

In Uttar Pradesh all the reservoirs are multipurpose. According to Directorate of Fisheries, Uttar Pradesh, the state is having 2.83 and 1.62-lakh hectares area under large and small reservoirs respectively. Sugunan (1995) estimates 31 small, 13 medium sized and one large reservoir in the state having fish yield of 14.60, 7.17 and 1.07 kg/ha/yr respectively. Punjab has less potential for reservoir fisheries. There are four dams. The naturally available fish species are the major carps, common carps, *Tor putitora, Labeo dero, Labeo bata, Mystus seenghala, Synothorax* etc. (IIMA, 1985).

## Orissa

In Orissa the total reservoir area is 99,061 ha (IIMA, 1985). Sreenivasan (1998) estimated about 1,19,403 ha large, 9,246 ha medium and 650,47 ha small sized reservoirs. According to Sugunan (1995), the state is having a total of 1442 reservoirs with an area of 1,98,198 ha.

## West Bengal

It is an important maritime state on the east coast of India having only 7 reservoirs, out of which two are major ones (IIMA, 1985). According to Sugunan (1995), the state is having six reservoirs with combined water spread area of 15, 732 hectares. No correct data on fish yield, number of reservoirs and stocking is available as the pond culture practices are more dominant than the reservoir fisheries. In Bihar, total reservoir area is about 50,000 hectares. There are 350 fishermen's cooperative societies in the state with total membership of 27, 142, which are under control of state cooperative department. The functioning of these societies are not satisfactory and are dominated by influential persons who exploit the members.

## Madhya Pradesh

It is purely an inland fish producing state. According to Sugunan (1995), the total number of reservoirs in the state is 32 with water-spread area of 460384 hectares. IIMA (1985) reported the large and small reservoirs with a total water spread area of 2.5 lakh hectares. The fish production for the period of 1974-75 to 1979-80 from selected reservoirs was varied from 0.20 to 87 kg/ha/yr (IIMA, 1985). According to Sreenivasan (1998), state is having 1.17-lakh hectare of reservoirs yielding an average fish catch of 26.46 kg/ha. George (2001) studied Gandhisagar, Bargi, Tawa, Barna and Halali reservoirs to compare the efficiency of the management measures followed. In the present scenario, ownership of these water bodies by apex federations of primary fishermen cooperatives seems to be the best option for their management for their management and sustainable exploitation.

## Gujarat

The Gujarat has about 0.14 million ha. area under reservoirs (IIMA, 1985).According to State Fisheries Department of Gujarat, the total area under reservoirs in the state is 286230 hectares and more than 95 per cent (in the number) of these man made lakes belong to the small category, although they form only 29 per cent of the total area. Anon (1980, 1983, 1984b) furnished the information on few reservoirs in the state.

### Maharashtra

It is one of the largest states in the country in population and geographical area having a number of rivers like the Godavari, the Bhima, the Krishna, the Narmada, the Tapti and other several rivers and their tributaries having a total of 1600 km of river length. Sugunan (1995) mentions the total reservoir area in the state is 2,73,750 ha. However, according to Sreenivasan (1998), Maharashtra is endowed with 1,79,430 ha. of reservoir area and the state produced 7.83 kg/ha fish from its reservoirs. However IIMA (1985) worked out 1,05,202 ha. reservoir area comprising 72 reservoirs. Pathak (1990) is of the view that, the area under reservoirs in the state of 20 ha and above estimated to about 2,36,157 hectares. No limnological studies or survey of fish fauna has been carried out so as to know the productivity of water and fish faunal diversity. However, Sugunan (1995), Sreenivasan (1991), Valsangkar (1980 and 93), Sakhare (1999, 2001, 2003, and 2005), Sakhare and Joshi (2002, 2003, and 2004), IIMA (1985), and Desai (1980) documented information on fisheries and/or ecology of some of the reservoirs in the state.

## Crafts and Gears Used for Fishing

The success of reservoir fisheries management depends on the quality of fish seed, proper stocking, exploitation of fish with appropriate crafts and gears, providing ice and cold storage facilities for better marketing of fish. Crafts used in the Indian reservoirs are of primitive type, and no modern crafts have been introduced for fishing. The craft called *'Nav'* is common in reservoirs of Maharashtra.This craft is nothing but a platform of 6x3 feet size with a depth of 4-10 inches is prepared from the thermocoel and covered by a plastic covering. The cost of one craft varies from 400-600 in the local markets. Generally the motorboats are not used because of heavy expenditure and other mechanical problems. The cooperative societies engaged in fishing are not in a position to provide loans or subsidy for the purchase of wooden boats/ motorboats; because the members have not returned the loans already taken. Devi (1997) accounted traditional catamaran improved with thermocoel and coracles from Ibrahimbagh and

**Figure 11.1: Fishing in Progress**

Shathamraj reservoirs of Hyderabad.There are so many drawbacks of such type of crafts such as

1. They are not suitable to carry large nets, having heavy weight
2. When fishermen get large quantities of fish catches, it become difficult to carry them to the landing centers.

Gears used for the fishery are mostly the surface gill-nets. However, cast-nets and hook and lines are also used. The cost of net varies according to the quality of nylon used for their fabrication. Good quality nets are prepared from the Garware nylon, which is considered good quality nylon yarn. Khan *et al.* (1991) mentioned that the gill-nets used in almost all reservoirs require improvements in respect of design, rigging and other parameters like mesh size, and twin size in order to increase its efficiency.

## Seed Stocking

According to Raghavachari (1983) stocking is not being done systematically in Indian reservoirs. There is a shortage of seeds and fingerlings. The size of fingerlings stocked has a direct bearing on the production at the later stage. Many reservoirs do not have a fish seed farm at the site. Even if there is one, the seed farm is unable to meet the needs of the reservoir. Raghavachari and Surychandra Rao (1984) and Srivastava (1985) recommended the following stocking rates:

Large reservoirs (above 5000 ha)–200 fingerlings/ha

Medium reservoirs (1000-5000 ha)–400 fingerlings/ha

Small reservoirs (less than 1000 ha)–1000 fingerlings/ha.

Raghavachari and Suryachandra Rao (1984) further mentions that the stocking should be done in one size in the form of advanced fingerlings with size of 12.5 cm. Paul and Sugunan (1990) reported the stocking programme in reservoirs had been generally devoid of any ration are behind them and consequently, no desirable impact was visible in most of the cases. They also stated that in many reservoirs stocking programme on scientific basis were impaired due to lack of seed availability. They suggested that the matters regarding selection of stocking materials, size of fingerlings and the stocking rate should be decided after taking into account the relevant ecological data. Selvaraj and Murugesan (1990) reported the factors responsible for low yield in Aliyar reservoir of Tamil Nadu, one of the main factor was stocking. The size of fry stocked was usually so small that they fall an easy prey to the predatory fishes. Even when the seed of major carps was produced in the fish farm through hypophysation, the size of seed stocked remained very small leading to their high mortality and subsequent poor yield and low contribution by the major carps in the landings. Sinha (1998) recommended a rate of 250 fingerlings per ha for reservoirs without catfishes and 600 fingerlings (6″ in length) with rich catfish population. But, for new reservoirs, the stocking rate should be at a higher level (1000 kg/ha). Irrational stocking of fingerlings has a deleterious effect on reservoir fisheries. However, a rational approach for formulating stocking policy is through estimation of potential fish yield of the reservoir and adjustment of stocking rate in such a manner as to obtain the yield close to the productivity potential. Stocking of economically important, fast growing fishes from outside is aimed at colonizing all the diverse niche of the biotope for harvesting maximum sustainable crop from them. Central Inland Fisheries Research Institute (1997) under the All India Coordinated Projects on Ecology and Fisheries of Freshwater Reservoirs carried out investigations on Bhatghar reservoir of Pune district of Maharashtra.The investigation recommended the need for regular stocking of major carp fingerlings. Sakhare (2003) reported autostocking in Yeldari reservoir of Maharashtra.The similar type of observations were also made by Devi (1997) in Ibrahimbagh and Shathamraj reservoirs of Hyderabad.

According to Sreenivasan (1989) when existing species are inferior, it would be necessary to add superior varieties. Species esteemed by consumers (taste, texture, flavour) have to be introduced. For Indian reservoirs, the following species are suitable–*Catla, Rohu, Mrigal, Labeo calbasu. Cirrhinus cirrhosa, Mahseers, Labeo bata,* and in some cases *L. fimbriatus.* Among exotics the common carp (*Cyprinus carpio*), Bangkok variety is suitable only for certain reservoirs. The Chinese carps, grass carp (*Ctenopharyngodon idella*) and silver carp (*Hypopthalmicthys molitrix*) have not been viewed with favour and have raised a dust of controversy. The example of role of silver carp in Gobindsagar is cited to prove the adverse effect of this species in reservoirs. These species found in Gobindsagar conditions similar to its native China and spawned prolifically and continuously. The same situation is not likely to occur in reservoirs warmer than this and further down south. *Tilapia mossambica* is not a recommended species for reservoirs with good carp fishery. Pathak (1990) suggested fingerlings of 100-150 mm size of commercially important species (Catla, Rohu, Mrigal and Common carp) need to be stocked in reservoirs, which would withstand the pressure of predation to some extent. He further recommended to stock large, medium and small sized reservoirs at the rate of 500, 1000 and 1500 fingerlings/ha respectively. He also experienced that whichever reservoir was stocked properly in the past, they gave a very good fish production.

## Fish Fauna and Fish Production

Depending on the available literature, the fish production from small, medium and large sized reservoirs in India varies from 0.52 to 912.7 kg/ha/yr. The highest rate of fish production was recorded from small reservoirs and lowest from large sized reservoirs. Sugunan and Yadava (1992) Sakhare (1999, 2001, and 2003), Sakhare and Joshi (2002 and 2003), Valsangkar (1993), Singh (2001), Piska *et al.* (2000) and many other workers reported the fish fauna of various Indian reservoirs. The common fishes recorded from Indian reservoirs are listed in Table 11.2.The checklist is prepared with the help of research papers published by various scientists and institutions in country.

**Table 11.2: Fish Fauna of Indian Reservoirs**

**Order: Clupiformes**–*Notopterus notopterus, N.chitala*

**Order: Perciformes**–*Chanda nama, C.ranga*

**Order: Cypriniformes**–*Salmostoma boepia, S.untrahi, Rasbora elanga, Rasbora danoconous, Danio equipinnatuus, D.devario, Catla catla, Labeo rohita, L.calbasu, L.fimbratus, L.boggut, L.porcellus, L.sindensis, Cirrhina mrigala, Cirrhina fulungee, Cyprinus carpio, Garra lamta, Hypopthalmicthyes molitrix, Osteobrama vigorsii, O.neilli, O.cotio cotio, Rohtee ogilbii, Tor khudree, T.mussal, Puntius sophore, P.kolus, P.ticto ticto, P.sarana, P.jerdoni, P.amphibious, P.subnastutus, Nemachilus evazardi, Nemachilus denisonii, Nimachilus rueprel.*

**Order: Siluriformes**–*Ompak bimaculatus, O.pabda, Silonia silondia, Heteropnestus fossilis, Wallago attu, Pseudotropius etherionoides, Mystus seenghala, M.cavasisus, M.vittatus, M.gulio, M.aor, M.tengra, Xenentodon cancilla, Rhinomugil corsula, Channa marulius, C.gachua, C.striatus, Bagarius bagarius, Pangasius pangasius, Clarias batrachus, Glossogobius giuris giuris*

## Declaration of Closed Season and Poaching

Raha and Sarkar (1980) reported the declaration of closed season during monsoon season (from April to June), when no fishing is allowed in the reservoir. Anybody found fishing in the reservoir during the closed season or without licence during the fishing season is liable to be prosecuted before the court under Indian Fisheries Act.

Poaching is also a notable thing in many reservoirs as there is no sufficient staff to prevent it. George (2001) also mentioned poaching as a major problem in Gobindsagar, Bargi and Tawa reservoirs of Madhaya Pradesh. It is due to the lack of manpower, heavy costs, vastness of the area, hostile terrain and absence of vehicles. According to Raghavacharis and Rao (1984) in some Indian reservoirs poaching yields greater production than the legal fishing and it is difficult to check poaching because of vastness of the reservoir and the terrain around the reservoirs. Negligible poaching was observed in Gobindsagar and Pong dam; they are also of the view that due to high cost, an official agency alone cannot check the poaching. Sugunan and Yadava (1991) mentioned necessity of boat which will serve to transport fish from gill net points to the landing center and the same boat can be used for patrolling the Nongmahir reservoir (Meghalaya) to ward off poachers and to check undesired fishing activities.

## Deforestation of Standing Trees

Deforestation of standing trees in many reservoirs has always remained neglected. Ringangaonkar and Bhrushundi (1980) described the manual saw to cut down submerged standing trees in Nalganga reservoir. Trees from the area which were exposed during summer months when water level of reservoir goes down. In case of trees in deeper waters where saw cannot reach, the branches are cut down. Applying forces for cutting by a one person also cut standing trees in shallow water. But to cut the tree under water the saw should reach the bottom of the tree-ordinary saw operated by two persons was not sufficiently long to reach the bottom. The saw length was increased by fixing long handles According to Jhingran (1985) the problem of clearance of submerged obstruction was squarely tackled in a number of small reservoirs constructed in Madhya Pradesh, Chillar and Benisagar being well known examples where from the standing on reservoir bed were cut and the wood auctioned. Even after the reservoir is filled it is possible to clear the obstructions at least in the reservoir margins but at that stage the work can be taken up only in the summer months when the water level goes down and the shore areas are exposed. He further mentioned that the presence of aquatic vegetation not only hampers fishing but also suppresses growth of benthic fauna as it occupies the productive zones. A machine has been developed at the Central Institute of Fisheries Technology for eradication of submerged weeds, which consists of a rotating fork arrangement driven by an engine, the rake and engine having been installed in a self-propelled pantoon. The cost of clearing works out to be Rs. 100 per hectare. A clearance of about 80 per cent is obtained. About 1 to 1.25 ha per day of 8 hours operation can be cleared.

**Table 11.3: Few Important Reservoirs in India**

| *Sl.No.* | *Reservoir* | *State* | *Area (ha)* | *Year of Construction* | *River* |
|---|---|---|---|---|---|
| 1. | Mettur | Tamil Nadu | 15346 | 1934 | Cauvery |
| 2. | Bhavanisagar | Tamil Nadu | 7886 | 1953 | Bhavani |
| 3. | Sathanur | Tamil Nadu | 2010 | 1957 | Ponniar |
| 4. | Stanley | Tamil Nadu | 15346 | 1939 | Cauvery |
| 5. | Malampuzha | Kerala | 2313 | — | Bharatapuzha |
| 6. | Idukki | Kerala | 6160 | 1973 | Periyar |
| 7. | Parappar | Kerala | 2590 | 1987 | Kallada |
| 8. | Chulliar | Kerala | 165 | 1970 | Chulliar |
| 9. | Tungabhadra | Karnataka | 37814 | — | Tungabhadra |
| 10. | Markonahalli | Karnataka | 1336 | 1939 | Shimsha |
| 11. | Hemavathy | Karnataka | 9162 | 1981 | Hemavathy |
| 12. | Vanivilas Sagar | Karnataka | 8640 | 1901 | Vedavati |
| 13. | Krishnarajsagar | Karnataka | 13200 | 1932 | Cauvery |
| 14. | Nagargunasagar | Andhra Pradesh | 28474 | 1967 | Krishna |
| 15. | Yerrakalava | Andhra Pradesh | 1550 | 1988 | Yerrakalava & Jalleru |
| 16. | Hussainsagar | Andhra Pradesh | 446 | 1850 | Kukatpally |
| 17. | Yeldari | Maharashtra | 9472 | 1962 | Purna |

*Contd...*

**Table 11.3–Contd...**

| Sl.No. | Reservoir | State | Area (ha) | Year of Construction | River |
|---|---|---|---|---|---|
| 18. | Nathsagar | Maharashtra | 39777 | 1975 | Godavari |
| 19. | Ujini | Maharashtra | 29000 | 1978 | Bhima |
| 20. | Siddeshwar | Maharashtra | 3360 | 1962 | Purna |
| 21. | Palas Nilegaon | Maharashtra | 206 | 1979 | Bori |
| 22. | Bori | Maharashtra | 746 | 1966 | Bori |
| 23. | Manjara | Maharashtra | 4590 | 1981 | Manjara |
| 24. | Bhatghar | Maharashtra | 2800 | 1928 | Yelwandi |
| 25. | Gangapur | Maharashtra | 2400 | 1973 | Godavari |
| 26. | Shivajisagar | Maharashtra | 11535 | 1961 | Koyana |
| 27. | Girna | Maharashtra | 5420 | 1970 | Girna and Panzan |
| 28. | Jawalgaon | Maharashtra | 858 | 1977 | Nagzari |
| 29. | Hingni pangaon | Maharashtra | 1006 | 1976 | Bhogawati |
| 30. | Ekruk | Maharashtra | 1842 | 1871 | Aghila |
| 31. | Khadakwasla | Maharashtra | 1472 | 1967 | Mutha |
| 32. | Mula | Maharashtra | 5324 | 1978 | Mula |
| 33. | Bhandardara | Maharashtra | 1822 | 1926 | Pravara |
| 34. | Dhom | Maharashtra | 2308 | 1977 | Krishna |
| 35. | Manar | Maharashtra | 2460 | 1964 | Manar |
| 36. | Kodar | Madhya Pradesh | 3583 | 1981 | Kodarnala |
| 37. | Gandhisagar | Madhya Pradesh | 66000 | 1960 | Chambal |
| 38. | Ravishankar-sagar | Madhya Pradesh | 9540 | 1978 | Mahanadi |
| 39. | Govindgarh | Madhya Pradesh | 307 | 1916 | Bichia |
| 40. | Tawa | Madhya Pradesh | 20055 | — | — |
| 41. | Bargi | Madhya Pradesh | 27296 | — | — |
| 42. | Halali | Madhya Pradesh | 7712 | — | — |
| 43. | Hirakud | Orissa | 71963 | — | Mahanadi |
| 44. | Rangali | Orissa | 37840 | — | Brahmani |
| 45. | Getalsud | Bihar | 3459 | 1971 | Subarnarekha |
| 46. | Konar | Bihar | 2590 | 1955 | Konar |
| 47. | Maithon | Bihar | 11490 | 1957 | — |

*Contd...*

**Table 11.3–Contd...**

| Sl.No. | Reservoir | State | Area (ha) | Year of Construction | River |
|---|---|---|---|---|---|
| 48. | Panchet | Bihar | 7511 | 1958 | Damodar |
| 49. | Badua | Bihar | 880 | 1963 | Badua |
| 50. | Nalkari | Bihar | 991 | 1967 | Nalkari |
| 51. | Rihand | Uttar Pradesh | 46538 | 1962 | Rihand |
| 52. | Baghla | Uttar Pradesh | 250 | 1952 | Barica |
| 53. | Nanaksagar | Uttar Pradesh | 4662 | 1957 | Chuka |
| 54. | Gobindsagar | Himachal Pradesh | 16867 | 1959 | Sutlej |
| 55. | Pong | Himachal Pradesh | 24629 | 1974 | Beas |
| 56. | Jaismand | Rajasthan | 7286 | 1730 | — |
| 57. | Ramgarh | Rajasthan | 1260 | 1903 | Banganga |
| 58. | Chhparwara | Rajasthan | 200 | 1894 | Mavshi |
| 59. | Mahibajaj | Rajasthan | 13500 | — | — |
| 60. | Kadana | Rajasthan | 9000 | — | — |
| 61. | Gudha | Rajasthan | 1928 | — | Mej |
| 62. | Keetham | Rajasthan | 306 | 1925 | Jamuna |
| 63. | Vallabhsagar | Gujarat | 52000 | 1971 | Tapti |
| 64. | Kadana | Gujarat | 16600 | 1979 | — |
| 65. | Panam | Gujarat | 8980 | 1977 | — |
| 66. | Gumti | Tripura | 4500 | 1976 | Gumti |
| 67. | Nongmahir | Meghalaya | 70 | 1979 | Umiam |
| 68. | Khandong | Assam and Meghalaya | 12950 | 1986 | Kapili |
| 69. | Kyrdemkulai | Meghalaya | 90 | — | Umiam |
| 70. | Umrong | Assam | 5550 | 1986 | Kapili |

## REFERENCES

Anon, 1980. *Proceedings of the 6th workshop of All India Coordinated Project on Ecology and Fisheries of Freshwater Reservoirs*, Simla 25–26 November, Central Inland Fisheries Research Institute, Barrackpore, West Bengal, pp 121.

Anon, 1983. *Proceedings of the 7th workshop of All India Coordinated Project on Ecology and Fisheries of Freshwater Reservoirs*,

Barrackpore, 9–10 March, Central Inland Fisheries Research Institute, Barrackpore, West Bengal, pp 148.

Anon, 1984b. New low cost coracle for fishermen. *CIFRI Newsletter*, 7(1 and 2): 3–4.

Brown, L., 1997. Facing the prospect of food scarcity. In L. Brown (Ed.) *State of the World*. Washington, DC: Worldwatch. pp. 23–41.

CIFRI. 1997. *Ecology and Fisheries of Bhatghar Reservoir*. Golden Jubilee Spl.Bulletin No. 73.

CIFRI. 1998. Ecological and fisheries status of reservoirs in India: A recent survey. In: Das, Manas Kr. (ed.). *The Inland Fisheries News*, Newsletter of CIFRI, 3(1): 1–2.

Costa-Percy, B.A., 1998. Constraints to the sustainability of cage aquaculture for resettlement from hydropower dams in Asia: An Indonesian case study. *J.Env. and Dev*. Paper available on internet.

Costa-Percy, B.A. and Soemarwoto, O., 1987. Proliferation of Asian reservoirs: The need for integrated management. Naga, The ICLARM quarterly, 10, 9–10.

Devi, Sarla, B., 1997. Present status, potentialities, management and economics of fisheries of two minor reservoirs of Hyderabad *Ph.D. Thesis*, Osmania University, Hyderabad.

De Silva, S.S., 1988. The reservoir fishery of Asia. In: S.S. De Silva (ed.) *Reservoir Fishery Management and Development in Asia*. International Development Research Centre, Ottawa, Canada, pp. 19–28.

De Silva, S.S., 1992. *Reservoir fisheries of Asia*. International Development Research Centre, Ottawa, Canada.

Dixtulu, J.V.H., 1999. Render Indian Reservoirs Sustainably Fishful. *Fishing Chimes*, 19(2): 5 –7.

Dwivedi, S.N., 1980. Reservoir fisheries for rural development : New polices and technologies. *India Today and Tomorrow*, 8(4): 156–159.

George, Ninan, 2001. Fishery management practices of the major reservoirs in Madhya Pradesh. Paper presented during *Nat.*

*Sem. on Riverine and Reservoir Fisheries: Challenges and Strategies*, May 23 to 24, Cochin.

Jaya Raju, P.B., 2000. Ecological wonders of India: (Ecological degradation due to intensive fish culture practices in lake Kolleru, Andhra Pradesh, (India). *Hydrosphere*, No. 6. IAAB, Hyderabad, pp. 40–46.

Jhingran Arun, G. and V.V. Sugunan, 1990. General guidelines and planning criteria for small reservoir fisheries management. p.18. In: Jhingran Arun G. and V.K. Unnithan (eds). *Reservoir Fisheries in India*.Proc.of the Nat. workshop on Reservoir Fisheries, 3–4 January. Special Publication 3, Asian Fisheries Society, Indian Branch, Mangalore, India

Jhingran, V.G., 1985. *Fish and Fisheries of India*. Hindustan Publishing Corporation (India), New Delhi, p. 106, 171, 191.

Mc Cully, P., 1996. *Silenced Rivers: The Ecology and Policies for Large Dams*. Zed Books, London.

Mohan Braj, 2001. Economic Status of Fishermen of Malampuzha Reservoir. Paper presented during the *Nat. Sem. on Riverine and Reservoir Fisheries: Challenges and Strategies*, May 23 to 24, Cochin.

Pathak, S.C., 1990. Harnessing reservoirs for increasing fish production. p 9–12. In: Jhingran Arun G. and V. K. Unnithan (eds). *Reservoir Fisheries in India*. Proc. of Nat. workshop on Reservoir Fisheries, 3–4 January, Spl. Publ. 3. Asian Fisheries Society, Indian Branch, Mangalore, India.

Piska, Ravishankar, 1999. *Fisheries and Aquaculture*. Lahari Publications, Hyderabad.

Piska, Ravishankar and K. Divakara Chary, 2000. Impact of trophic nature of reservoir on the reproductive capacity of catfish *Mystus bleekeri* (Day). *Ecol. Env. and Cons.*, 6(4): 447–452.

Postel, S., Daily, G., and Ehrlich, P., 1996. Human appropriation of renewable freshwater. *Science*, 271: 785–788.

Raha, K. and M.K. Sarkar, 1980. Salient features of the Gomti reservoir in Tripura. *India Today and Tomorrow*, 8(4): 178–180.

Raghavachari, M., 1983. Reservoir Fishes in India: Some Issues and Problems. In: U.K. Srivastava and M. Dharma Reddy (eds).

*Fisheries Management in India: Some Aspects of Policy Management*. Concept Publishing Company, New Delhi, pp. 459–461.

Raghavachari, M. and Rao, S.S., 1984. Development of reservoir fisheries in India: some issues and recommendations. 213–250. In: Srivastava, K. and S. Vathsala (Eds). *Strategy for Development of Inland Fishery Resources In India: Key Issues in Production and Marketing*. Proc. of Nat. Workshop on Dev. of Inland Fish resources held on November 1–3, 1983 at Ahmedabad.

Ringangaonkar, A.M.and M.G. Bhrushundi, 1980. Deforestation of Reservoir Basin : A new indigenous manual device. *India Today and Tomorrow*, 8(4): 172.

Sakhare, V.B., 1999. Fisheries of Yeldari Reservoir, Maharashtra. *Fishing Chimes*, 19(8): 45–47.

Sakhare, V.B., 2001. Ichthyofauna of Jawalgaon reservoir in Solapur district of Maharashtra. *J. Aqua. Biol.*, 16(1 and 2): 31–33.

Sakhare, V.B., 2001. Reservoir fisheries in Solapur district of Maharashtra. *Fishing Chimes*, 21(5): 29–30.

Sakhare, V.B., 2003. Studies on some aspects of fisheries management of Yeldari reservoir, Maharashtra. *Ph.D. Thesis*, Swami Ramanand Teerth Marathwada University, Nanded (Maharashtra).

Sakhare, V.B., 2005. Water quality of Hingni (Pangaon) reservoir and its significance to fisheries. In: *Advances in Limnology*, (Ed.) S.R.Mishra, Dhar. Daya Publishing House, New Delhi, pp. 231–235.

Sakhare, V.B., 2005. Ecology and fisheries of Manjara reservoir in Maharashtra. *Fishing Chimes*, 25(3): 42–45.

Sakhare, V.B. and P.K.Joshi, 2002. Ecology and ichthyofauna of Bori reservoir in Maharashtra. *Fishing Chimes*, 22(4): 40–41.

Sakhare, V.B. and P.K. Joshi, 2002. Ecology of Palas-Nilegaon reservoir in Osmanabad district, Maharashtra. *J. Aqua. Biol.*, 18(2): 17–22.

Sakhare, V.B. and P.K. Joshi, 2003. Reservoir fishery potential of Parbhani district of Maharashtra. *Fishing Chimes*, 23(5): 13–16.

Sakhare, V.B. and P.K. Joshi, 2004. Present status of reservoir fisheries in Maharashtra. *Fishing Chimes*, 24(8): 56–60.

Selvaraj C, and K. Murugeson, 1990. Management techniques adopted for achieving a record fish yield from Aliyar reservoir, Tamil Nadu. pp. 86–96. In: Jhingran, Arun G. and V. K. Unnithan (eds). *Reservoir Fisheries in India.* Proc. of the Nat. workshop on Reservoir Fisherie , 3–4 January. Spl. Pub. 3, Asian Fisheries Society, Indian Branch, Mangalore, India.

Shetty, H.P.C., 1990. Forward. Proc. of Nat. Workshop on Reservoir Fisheries, 3–4 January. Spl. Publ. Asian Fisheries Society, Indian Branch, Mangalore, India

Singh, Gurucharan, 2001. Status of Development of fisheries of Pong Reservoir (Himachal Pradesh). *Fishing Chimes,* 21(1): 88–90.

Sinha, M., 1998. Policy Options for Integrated Development of Reservoir Fisheries: From Production to Marketing. *Fishing Chimes,* 18(1): 54–59.

Sreenivasan, A., 1989. Principles of ecology and fisheries management of man made lakes. In: *Conservation and Management of Inland Capture Fisheries Resources of India,* (Eds.) Arun G. Jhingran and V.V. Sugunan, Inland Fishery Society of India, Barrackpore, pp. 96–105.

Sreenivasan, A., 1998. Integrated Development of Reservoir Fisheries of India: Production to Marketing. *Fishing Chimes,* 18(1): 60–63.

Srivastava, U.K., D.K. Desai, V. K. Gupta, S.S. Rao, G.S. Gupta, M. Raghavachari and S. Vatsala. 1985. *Inland Fish Marketing in India: Reservoir Fisheries,* Vol. 4 (A and B). Concept Publishing Co., New Delhi, pp (A) 403 and (B) 1184.

Sugunan, V.V. and Y.S. Yadava, 1991b. Feasibility studies for fisheries development of Nongamahir reservoir. CIFRI, Barrackpore, pp. 30.

Sugunan, V.V., 1995. *Reservoir Fisheries of India.* FAO Fisheries Tech. Report 345. Daya Publishing House, Delhi.

Sugunan, V.V., 1990. Reservoir fisheries management. In: Sugunan V.V. and U.Bhowmick (eds) *Technologies for Inland Fisheries Development.* Central Inland Fisheries Research Institute, Barrackpore, India, pp. 153–164.

Sukumaran, P.K. and A.K. Das, 2001. Status of plankton in some freshwater reservoirs of Karnataka. Paper presented during the

*Nat. Sem. on Riverine and Reservoir Fisheries: Challenges and Strategies*, May 23 to 24, Cochin.

Valsangkar, S.V., 1980. Economic rehabilitation of fishermen in Yeldari reservoir. *India Today and Tomorrow*, 8(4): 162–163.

Valsangkar, S.V., 1993. Mahseer fisheries of Koyana river (Shivajisagar) in Maharashtra: Scrap to Bonanza. *Fishing Chimes*, 12(10): 15–19.

## Chapter 12

# Fisheries Institutions in India

The fisheries sector in India has experienced remarkable growth and development over the last decade. It has been recognized as a powerful income and employment generator as it stimulates growth of a number of subsidiary industries and is a source of cheap and nutritious food, at the same time it is a tool of livelihood for a large section of economically backward population of the country. More than 7 million people in the country depend on fisheries and aquaculture for their livelihood. Indian fisheries is an important component of the global fisheries, with India being the third largest producer of fish in the world and second in inland fish production. India's share in the global fish production has increased from 3.2 per cent in 1981 to 4.5 at present. Fish production in the country has been showing an increasing trend and has reached a record level of 5.65 million tones in 1999-2000.The estimated fish catch during year 2000-01 is about 5.95 million tones. This continued progress is greatly dependent on the close cooperation among scientists who work in different aspects of the fisheries industry to tackle the constraints confronting it.

The fisheries education in India started a long back in order to fulfill the challenging demand of food security for exponentially increasing population. Some of the institutions and colleges conducting research, training and various courses in fisheries and aquaculture are:

## Central Inland Fisheries Research Institute (C.I.F.R.I.), Barrackpore (West Bengal)

It was formed as central Inland Fisheries Research Station at Kolkata on 7th March 1947. By the year 1959, the station acquired the status of an institute (CIFRI) and moved to its own building at Barrackpore, West Bengal. The institute is having different divisions mainly working on resource assessment, riverine fisheries, reservoir fisheries, estuarine fisheries, flood plain, wetland fisheries, environmental management, fish health and hilsa fisheries.

The present mandate of the institute involves the study of fish population dynamics of exploitable inland open water ecosystems, causes and effects of population in open water fishery resources and impact of population in open water fishery resources and impact of river valley projects on the ecology of river basin. Besides providing consultancy services the institute also conducts training, education and extension education programmes.

CIFRI has developed a technology like riverine carp spawn prospecting, carp seed production and composite fish culture which are readily accepted by fish farmers throughout the country. The division of CIFRI and its headquarters are:

1. Division of reservoir fisheries-Bangalore
2. Division of riverine fisheries-Allahabad
3. Division of Estuarine fisheries –Barrackpore
4. Division of wetland fisheries-Kolkata
5. Division of Environmental monitoring-Barrackpore
6. Division of Hilsa fisheries-Farraka/Vadodara
7. Division of Resource Assessment-Barrackpore
8. Division of Training and Extension-Barrackpore

## Central Institute of Fisheries Education (C.I.F.E.), Versova, Mumbai

The institute was established in year 1961.The mandate of the institute was to train the middle level fisheries officers by two-year course. During 1984, first batch of M.Sc. (Fisheries Management) was started. In year 1989, Government of India gives status of deemed university to this institute. CIFE was established by GOI with the assistance of Food and Agriculture Organization and United Nations development programme. The institute is a pioneer in offering fishery education and training in Asia. The subcentres of CIFE are:

## Inland Fisheries Training Centre (IFTC) Salt Lake, Kolkata

Earlier this center was at Barrackpore but during year 1998 it moved to Salt Lake (Kolkata). A certificate course in Inland fisheries development and administration is conducted at this centre. The course includes fish biology, classification and identification of fishes, inland fish culture, including induced breeding, riverine and estuarine fisheries, water soil and plankton analysis, conservation and management, marketing, commodity development, fisheries administration and extension.

## Aquaculture Research and Training Centre, Kakinada (Andhra Pradesh)

Every year the courses on freshwater fish seed production technology including carps and catfishes, freshwater aquaculture technology, Brakishwater aquaculture technology, Macrobrachium hatchery management and seed production technology and fisheries extension are conducted. The duration of the course varies from 45 days to 60 days.

## Inland Fisheries Operative Training Centre, Kathota Tal, P.O. Chinhat (Uttar Pradesh)

This centre was established mainly for the training of inland fisheries.

## Saline Ecosystem Aquaculture Research Centre, Rohtak (Haryana)

Plenty of area in state of Haryana is suitable for saline water prawn farming. This sub centre is conducting the training programme and research on saline water aquaculture. CIFE at its headquarter

Mumbai offer different courses at post-graduate and Ph.D. levels. The courses offered are:

M.F.Sc. (Fishery Management)

M.F.Sc. (Aquaculture)

M.F.Sc. (Processing technology)

M.F.Sc. (Inland Fisheries Administration and Management)

D.F.Sc. (Diploma in Fishery Science) is a two year's diploma course which is conducted for middle level state fisheries officers and private candidates from India and abroad. For the admission to M.F.Sc.courses, all India level entrance examination is conducted every year in the month of June by Indian Council of Agricultural Research, New Delhi. The eligibility for the PG programme is B.F.Sc.or Zoology as a subject at B.Sc. level with minimum 60 per cent marks. The notification for the examination comes in national newspapers. The Ph.D.programmes is also offered by institute in various aspects of freshwater and marine water fisheries. The all India level entrance examination is conducted for the admissions to Ph.D. course.

## Central Marine Fisheries Research Institute (C.M.F.R.I.), Cochin

Under the PG programme in mariculture, the institute organizes M.F.Sc. and Ph.D. programme. The course programme covers aspects of subject including dissertations for these programmes were affiliated to the Cochin University of Science and Technology and recently they were transferred to the Central Institute of Fisheries Education, Mumbai.

CMFRI has been offering need-based training within the preview of the institute's mandate. The training programmes include subjects such as fishery statistics, population dynamics, hatchery production of prawn seed, SCUBA dilling and seaweed culture.

The multidisciplinary research on capture and culture fisheries are conducted under eight divisions. These eight different divisions mainly works on fisheries resources assessment, pelagic fisheries, demersal fisheries, molluscan fisheries, crustacean fisheries, fishery environment management, physiology, nutrition and pathology, and socio-economic evaluation and technology transfer division.

## Central Institute of Freshwater Aquaculture (C.I.F.A.) Bhubaneshwar, Orissa

Foundation stone of this institute was laid by late Prime Minister Mrs.Indira Gandhi on 3rd January 1977. The mandate of institute are:

1. To conduct research more specifically in fish nutrition, physiology, genetics and pathology, pond environmental monitoring and aquaculture engineering for developing intensive and extensive warm freshwater fish farming system for commercially important fin fish and shell fish.
2. To conduct special training and extension programmes in freshwater aquaculture to enable in freshwater aquaculture economic utilization of the cultivated and cultivatable freshwater aquatic resources of the country.
3. To act as a nodal agency to provide scientific information and technology transfer for the freshwater aquaculture development.

The CIFA is engaged in research in different aspects of freshwater aquaculture like carp breeding and culture, breeding and culture of catfish and freshwater prawns, freshwater pearl culture, aquatic environment, fish genetics, fish physiology, fish nutrition, fish pathology, aquaculture engineering, aquaculture economics and statistics, and aquaculture extension. The institute acts as a nodal agency to provide scientific information for aquaculture development in the country. Advance training and extension services are also rendered with a view to transfer the know how on the recent developments in research activities in the sector.

CIFA is a regional lead centre on carp farming under the FAO/ UNDP Network of Aquaculture Centres in Asia-Pacific (NACA) as also the centre of advanced studies in freshwater aquaculture with degree programmes of M.F.Sc, and Ph.D. in collaboration with the Orissa University of Agriculture and Technology, Bhubaneshwar.

The training is offered to officials of the state fisheries departments, University teachers and researchers, KVK teachers, NGO officials and other engaged in R and D of freshwater aquaculture in the country. All the employed candidates should apply through proper channel with biodata. In case of others, direct applications are received. The training course is offered on Aquatic microbiology in freshwater aquaculture, cage culture of carps, advances in carp breeding, Magur (*Clarias*) breeding and culture, breeding and hatchery management in freshwater aquaculture and breeding and hatchery management of commercially important frog species. Institute is also having subcentres in various parts of country *i.e.*,

1. Wastewater Aquaculture Division, Rahara, West Bengal
2. Frog Breeding Culture Division, Bangalore, Karnataka.
3. Peninsular Aquaculture Division, Bangalore, Karnataka.
4. Prawn Breeding Centre, Kakinada, Andhra Pradesh
5. Krishi Vigyan Kendra/TTC, Kausalyaganga, Orissa.

## Central Institute of Brackishwater Aquaculture (C.I.B.A.), Chennai

The institute was established on 1$^{st}$ April 1986 during 7$^{th}$ five year plan as a part of the organization of the fisheries research institutes of the ICAR. The institute has the following mandates:

1. To conduct research leading to development of techno-economically viable and sustainable culture systems for finfish and shell fish in brakishwater.
2. To carry out research, more specially on nutrition and feed development, reproductive physiology, pathology, genetics, pond environment, aquaculture engineering and operational economics to provide the technology support for optimizing productivity and production.
3. To undertake transfer of technology through training, education and extension and extension programmes and to provide institutional consultancy services.

The research programmes of the institute were carried out on various aspects under crustacean culture, fish culture, technology improvement, aquaculture engineering, economics, extension and information divisions.

## National Research Centre on Coldwater Fisheries, Bhimtal (Uttar Pradesh)

In 1999 National Research Centre on Coldwater Fisheries was born as independent centre directly under ICAR control with headquarters at Haldwani in Uttar Pradesh. Subsequently from June 1997 the headquarter was shifted to Bhimtal, about 30 km from Haldwani.

Mandate of institute: Mandate is to promote and conduct research, training and extension in temperate aquaculture and open water bodies. To achieve this mandate the NRCCWF has the following functions:

1. Evaluate and assess the coldwater fishery resources in upland areas.
2. Develop strategies for their conservation and management.
3. Conduct research leading to development of suitable technology for forming of indigenous and exotic fish species in uplands.
4. Study the impact of environmental changes in the aquatic biodiversity in upland open waters.
5. Undertake transfer of technology through training, education, and extension programmes.
6. To provide consultancy services in different areas like coldwater development, aquatic ecology and environmental impact assessment.

Thus, all these fisheries institutions are serving for the development of Indian fisheries through the innovations and extension programmes.

## Chapter 13

# Culture and Breeding of Murrels

Murrels belonging to family channidae form an important group of food fishes in freshwater habitats. They are also known as 'snakeheads'. It is due to their snake like head. The various characteristics which make it a desirable cultivable species, include, rapid growth, ability to utilize atmospheric air for respiration in oxygen depleted water and high market value.

From the culture point of view, in India only three species are economically important, *i.e., Channa marulius* (Hamilton), *Channa punctatus* (Bloch) and *Channa striatus* (Bloch).

## *Channa marulius*

It is distributed in India, Bangladesh, Thailand and China. It inhabits lakes, reservoirs, and river but prefers deep clear stretches of water with sandy or rocky bottoms. It is commonly known as 'giant murrel'.

Body is elongated and subcylindrical anteriorly. Head depressed and covered with moderate sized scales. Snout somewhat obtuse. Mouth opening moderate, jaws equal. Teeth on jaws, vomer and palate. Eyes lateral in position and lie in anterior part of head. Accessory respiratory organs in the form of thin bony laminae

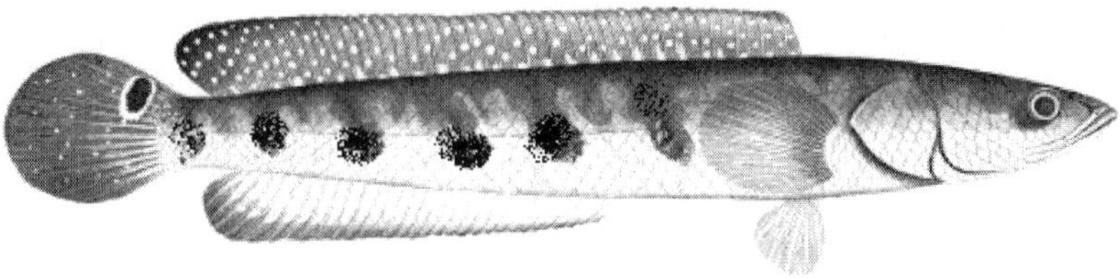

**Figure 13.1: *Channa marulius***

present in suprabranchial chamber. Dorsal fin long without spine. Both dorsal and anal fins free from caudal fin. Caudal fin rounded. Lateral line complete, which runs straight up to $17^{th}$ scale and then descends down. Colour varies with age and locality. Greyish green on dorsal side, and pale yellow below 4-5 black blotches located along lateral line. There is a large black spot with orange boundary at the base of caudal fin.

It grows up to 120cm in length. It attains maximum size of about 30cm at the end of first year.

Larval stages feeds on zooplankton including copepods and rotifers. It feeds more actively during August to November and January to March. Minimum feeding intensity is observed during spawning period. It is highly predaceous, even cannibalistic.

Fish attains maturity in second year of life. Matured male was reported when size was 300 mm in length and 235 gms in weight. While in female length was 320mm and weight was 252 gms. Fecundity in a length group of 500-820mm ranges from 2214-18475 eggs. The fertilized eggs measure about 1.9mm in diameter and are brownish in colour. It breeds from April to October in natural waters.

## *Channa punctatus*

It is distributed in India, Afganisthan, Pakistan, Nepal, Bangladesh and China. It is commonly found in freshwater ponds and rarely in flowing waters. It is commonly known as 'spotted murrel'.

Body is elongated, subcylindrical anteriorly. Head depressed and covered with large and irregular scales. Snout somewhat obtuse. Lower jaw longer. Teeth present on jaw, vomer and palatine.

Accessory respiratory organs in the form of thin bony laminae present in gill chamber. Dorsal fin without spine. Pectoral fins nearer to pelvic fins. Caudal fin rounded. Colour black or dark grey on dorsal side becoming light green or dull on ventral side. Several vertical bands from dorsal side of body pass downwardly to the middle of body. Fins dark grey and spotted. It attains maximum length of 25-30 cm.

It is carnivorous and predatory fish and feed upon aquatic insects, crustaceans, shrimps and molluscs.

It becomes mature in first year and breeds throughout the year with peak breeding before and during monsoon months. Fecundity ranges from 3000-26000 eggs. It lays eggs in nests in shallow marginal weeds. Parents guard the nest.

## *Channa striatus*

Distributed in India, Bangladesh, Sri Lanka, Vietnam, Cambodia, Indonesia, Thailand, Pakistan, Burma and China. It inhabits freshwater rivers, channels, ponds, lakes, paddy fields, tanks and reservoirs, but prefers stagnant muddy water, swamps and grassy tanks. It is commonly known as 'stripped murrel'.

Body elongated and rounded anteriorly. Head depressed and covered with plate like scales. Mouth opening wide, lower jaw longer. Teeth on lower jaw, vomer and palate. Gill openings wide. Accessory respiratory organs in the form of thin bony laminae present in supra branchial chamber. Dorsal fin without spine. Dorsal and anal fins free from caudal fin. Caudal fin rounded. The colour of fish varies with water. Colour is dark grey or blackish on dorsal side, yellowish white beneath. Grey or black transverse bands on thee sides of the abdomen. Ventral and anal fin greyish.

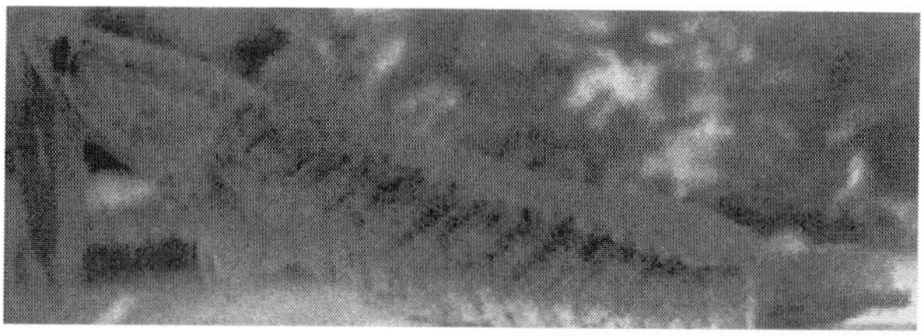

**Figure 13.2:** ***Channa striatus***

Hatchlings feed on protozoa and unicellular algae and early fry on crustaceans like *Daphnia* and *Cyclops*. Adult feeds on small fishes, aquatic insects and tadpoles.

It attains maximum length of 60-90cm.According to Day (1878), it may attain a maximum length of 994mm (3 feet) or slightly more. The growth attained in first and second year varies from 150-160mm and from 310-320 mm respectively.

It attains sexual maturity when two years old. It breeds in two times in a year. Eggs are laid in nests and both parents guard the nest. The eggs are amber coloured, round, non-adhesive and are found floating in a mass in the center of the nest. The average diameter of the egg is about 1.53mm.The fecundity has been observed between 4422 to 20,000 depending upon the size.

In composite culture, *C.striatus* is used as police fish to control the numbers of young tilapia. In developing countries like Sri Lanka, *C.striatus* is an important food fish, which makes a significant contribution to the reservoir fishery.

## Breeding

The breeding period of murrels in India is already discussed. The murrels are sexually diamorphic and in nature the males are sexually diamorphic and in nature the males are bestowed with a dominant instinct of parental care. The external morphological features such as bulging abdomen, round and reddish vent, broad slightly reddish genital papilla like structures, with blunt tip and reddish dot are used to assess maturity in female. Males do not ooze milt even when pressure is applied on the abdomen. The maturity of male can also be assessed by external examination of genital papilla, which becomes prominent. Breeding set consists of 2 male fish and 1 female fish. The females, during the breeding season show distended abdomen indicating mature ovaries. The male fish do not show any such character. In female, urogenital opening is circular in shape, which becomes suffused with blood during the breeding season. In males, it is elongated in contrast to that of the female.

Murrels shows parental care. The spawning takes place in nests built in shallow marginal areas with aquatic vegetation. Spawning lasts for 15-45 minutes and the eggs laid by the female in the nests are fertilized by the milt shed by the male. The fertilized eggs float in the center of the nest as a thin film. The eggs hatch out in 20-57 hours in temperatures ranging from 16-33 °C.The hatchlings measure about four days, when they reach a size of about 6.8mm, swim about freely. Both the sexes take part in caring for the newly hatched young for about 15-20 days. The larval development is completed in about 9 weeks. The larvae move to the bottom and show adult behaviour.

## Induced Breeding

For the cultivation of channa species a large number of quality seed is required, which is difficult to obtain naturally. Therefore, controlled breeding is desirable by inducing them to breed by hypophysation for which brood stock can be procured from the markets and maintained in ponds. Brooders ranging in weight from 0.5 to 1 kg are suitable for seed production by induced breeding.

Precollected pituitary glands of mature Indian major carps preserved in absolute alcohol are generally used for the induced breeding. For induced breeding, hypophysation with pituitaries at doses ranging from 40-80 mg/1000 gm female in two injections is found to be successful. Ovaprim at a dosage of 0.5 ml/kg body weight can be used for induced breeding of *C.striatus*.During spawning, the male bends its body close to the female and releases its milt and the eggs are fertilized externally. The fertilized eggs are usually buoyant in nature and adhere to each other forming an egg mass of 5-15cm diameter containing 2500to 4500 eggs. After hatchlings, the male moves around hatchlings and ventilates them with its pectoral fins. The fertilized eggs can be collected from the breeding pond using plastic basins and stocked in glass aquaria. Sometimes they are allowed to remain in the pond or tank as such, which helps the eggs to hatch out and the hatchlings to grow under parental care. The hatchlings float on the surface keeping the large ovoid yolk sac up the water. The yolk is absorbed by the end of the 2$^{nd}$ day in *C.punctatus* and by the end of 4$^{th}$ day in *C.striatus* and *C.marulius*.The air breathing habit develops around 20-25$^{th}$ day after hatching. To break the water surface, the larvae make an angle of 45° and return to the water bottom vertically down passively. The fry attain a body length of nearly 11mm when a regular practice of breaking the water surface begins.

Seed production in captivity and larval rearing of murrels have been accomplished experimentally but not on a large scale. In fishes under culture, phenomenon of cannibalism is usually associated with heterogeneous size variation, limited food availability, high stocking densities, limited refuge areas and light conditions. Prolonged cannibalism is observed in murrels during the fry and juvenile stages. The larger and healthier ones prey upon the smaller ones.

## Culture

The optimum temperature for the culture of murrels is in the range of 20-35°C.The common method of murrel culture is in earthen ponds ranging in size from 800 to 1600 $m^3$ and in depth from about 0.5 to 1m.To prevent fish escape, ponds can be provided with fine meshed wire. They can survive in hardy environments with low dissolved oxygen and high ammonia concentration and are often cultured in grow out ponds at stocking densities of 40-80 fish/$m^2$ with annual yields ranging from 7-15 mt/ha. Murrel fingerlings (about 4-5 cm) can be stocked in earthen ponds at the rate of 50 fingerlings/$m^2$ and reared for 6-10 months. The feed ratio can be given @ 5-10 per cent of the biomass daily.

The fingerling stage is the appropriate stage for release in culture pond. The stocking densities of murrel fingerlings with total length of 6-8 cm in grow out ponds range at 40-80 advanced fingerlings/ $m^2$ with annual yields of 7-15 t/ha. The recommended stocking rate of murrels in a grow out pond under Indian conditions is 20,000 fingerlings/ha for *C.marulius*, 30,000 fingerlings/ha for *C.striatus*, 40,000 fingerlings/ha for *C.punctatus* in semi-intensive culture operations and 40,000/ha, 60,000/ha and 80,000 /ha respectively in intensive culture.

## Feeding

The natural feed of the post larvae of *C.striatus* consists of cladocerans, copepods, phytoplanktonic organisms, rotifers and small aquatic insects. The hatchlings of murrels require intensive care. Rearing can be done with an artificial feed of boiled chicken egg yolk. Yolk of one egg is sufficient for 50,000 fingerlings/day. After seven days, the larvae are ready to feed on zooplankton. The nursery ponds with a low water column are suitable for rearing fingerlings of murrels. In traditional culture, murrels are fed with trash fish or cattle blood mixed with rice bran or wheat flour. Fry of *C.striatus* were trained to accept formulated feed consisting of fish paste, rice bran and wheat flour or fresh chicken intestine.

## REFERENCES

Dehadrai, P.V., Banerji, S.R., Nirmal, K.T. and Das, N.K., 1973. Sexual diamorphism in certain air breathing teleosts. *J. Inland Fish. Soc. India*, 5: 71–77.

Haniffa, M.A., Marimuthu, K. and Muruganandam, M., 2002. Mass breeding of the striped murrel induced through ovaprim. *Fishing Chimes*, 21(10 and 11): 53–54.

Marimuthu, K., M. Muruganandam and M.A. Haniffa, 1999. Rearing of larvae is an essential pre-requisite for murrel culture. *Fishing Chimes*, 18(12): 31–32.

Sampath, K., 1985. Food intake, conversion and surface activity as a function of density and water change frequency in an air breathing fish, *Channa striatus*. *Aquaculture*, 465: 201–213.

Selvaraj, S. and T. Francis, 2005. Breeding and culture of murrels in India. *Fishing Chimes*, 25(9): 19–22.

*Chapter 14*

# Culture and Breeding of Magur

Air breathing fishes such as catfishes and murrels are the economically important groups of fishes. They have special accessory respiratory organs and can be marketed alive. So they are known as 'live fishes'. The Asian catfish *'Clarias batrachus'* is popularly known as 'magur' or 'walking catfish'. Magur is widely distributed. It is found across southern Asia including Pakistan, Eastern India, Sri Lanka, Bangladesh, Myanmar, Thailand, Malaysia, Indonesia, Singapore, Borneo, Laos, Philippines, Combodia, Hongkong and China. It is a hardy fish, which can thrive where many other fish struggle to survive. In addition to lakes, reservoirs and lakes, they can be found in brakishwaters or warm, stagnant, often hypoxic waters such as muddy pond, canals, ditches, swamps and flooded prairies. They can remain dormant through periods of drought and go several months without eating. When they do eat, they consume a wide variety of prey. The fish is voracious, opportunistic feeders who are mainly active at night. They consume a wide variety of prey including eggs and larvae of other fishes, small fishes, a number of invertebrates including crustaceans and insects and sometimes plant materials. In densely populated drying pools, these fish become even more indiscriminate and quickly consume most other species present. The pectoral fins, one on each side, have rigid spine like elements.

To move outside of water, the fish uses these spines and flexes its body back and forth to 'walk'. They are well known for their ability to 'walk' on land for long distances, especially during or after rainfall. It is cultured in many countries like Thailand, Indonesia and Philippines, where very high production is reported up to 100 t/ha/year. Its culture is however yet to be picked up in many parts of India. The culture of magur has good prospects for developing domestic trade, because of its high market price, medicinal value, better taste, rich protein content and of less spines. It has present market rate of Rs. 50-75/kg and even fetches Rs.100-150/kg in metropolitans like Hyderabad.It is in great demand in the north-eastern part of India particularly in West Bengal, Assam, Orissa and Bihar for its high nutritional value. It contains higher percentage of protein and iron as compared to other edible species of fish. Its fat content is also very low and is therefore easily digestible so that it is very useful during convalescence. With most other it is a delicacy because of the characteristic aroma and softness of its flesh.

These fishes are not only able to thrive in water containing low dissolved oxygen but also they are extremely hardy with respect to all other environmental parameters and are suited to shallow and derelict waters as well as in saline impoundments having a salinity up to 10 ppt. It requires a relatively small area for culture. It can be stocked at higher density than any other culturable species and even in perennial stretch of water as the culture period is restricted to 5-6 months. Fish can live in water without oxygen and in water with a high content of $CO_2$.

## External Characters

Body is divided into head, trunk and tail. It is characterized by its spikeless dorsal fin, which extends all along the body; pectoral fin is inserted very low; anal fin is not confluent with caudal. Head is flat with four pairs barbels. The nasal ones reach up to the base of the occipital process and slightly longer than the mental barbels. Maxillary, the longest ones reach the base of pectorals. Mandibular barbels are a little shorter than the maxillary ones. Head bones are superficially exposed. Body is covered by scales and naked skin. Dendrtitic accessory branchial apparatus supplements gill respiration and hence fish can live for a very long period outside water. Eyes reduced and spiracles absent. Parietals, symplectics and sub-operculum absent. Tail is laterally compressed, diphycercal and

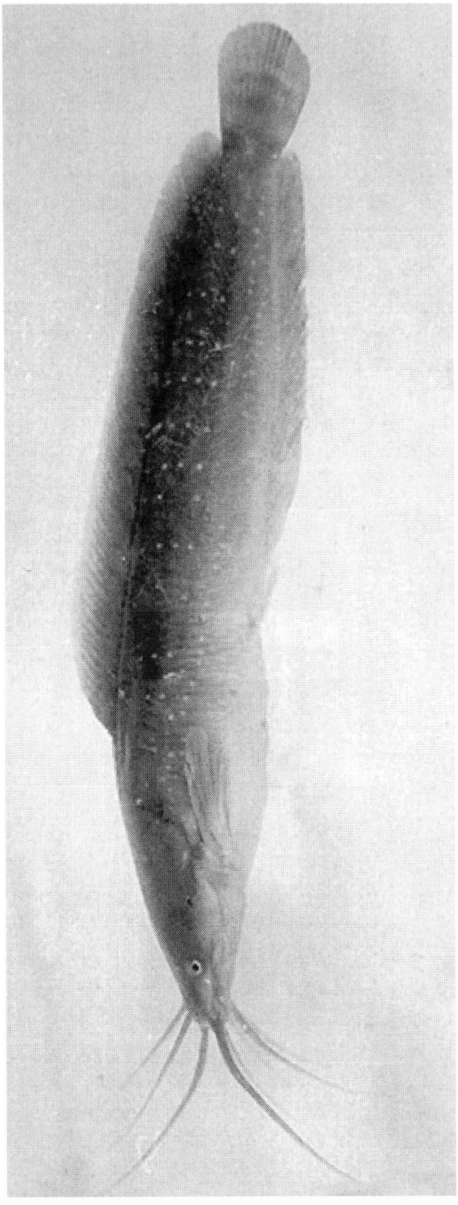

**Figure 14.1:** ***Clarias batrachus*** **(Magur)**

having rounded caudal fin. The pectoral fins cause painful wounds. They are placed very low along ventro-lateral angles of abdomen. All fins are usually with reddish margins. Body is tinged dark brownish superiorly and becoming lighter beneath. Dorsal, anal and caudal fins are fringed red.

## Food and Feeding Habit

It is carnivorous in habit. They prefer to eat decayed protein food and are considered as scavengers in pond. In cultured operation, they adapt themselves excellently to supplementary feeding with fishmeal, oil cake, rice bran, earthworms, insect larvae, silkworm pupae etc. Adult magur feeds on insect larvae, shrimps, worms, small fish and organic debris found in pond bottom. Fry feed on protozoans, small crustaceans, rotifers and phytoplanktons.

## Sexual Diamorphism

The sex of magur can be identified by the pointed anal papilla of the male and oval shaped papilla of the female. Sex identification is possible when the fish attains about 20 cm in length or of one-year age. The anal papilla of the adult male fish during breeding season remains pointed with reddish oval, swells up getting vascularised and the belly becomes distended.

## Fecundity and Egg Laying

Fecundity of magur is very low. Normally a fecundity of 3000-6000 numbers has been observed from fish ranging between 80 grams and 120 grams. Large sized fish above 130 grams are more fecund having fecundity to the tune of 10,000 numbers or more.

In nature the fish has been breed only in rainy season. The fish spawns in horizontal holes. It migrates to adjoining inundated paddy fields and spawn in the grassy bottom nests prepared by them.

The eggs are round and yellowish brown in colour. The size of egg ranges from 1.2 to 1.8mm.The eggs are adhesive in nature, cling to the grassy and rocky substrate of the nests/holes. Fish breeds at a temperature ranging from 25 to 30°C.Hatching take about 16-20 hours depending on the ambient temperature gradients.

## Parental Care

The male magur guards the nest to prevent devouring of the developing eggs by the predators (Tongsanga *et al.*, 1963).

## Larval Development

Yolk resorption usually takes 2-4 days depending on the water temperature.

## Method of Culture

Fingerlings (80-100 nos/kg) are collected from natural resources like paddy fields, canals etc. Because of high mortality rate, larger ones are stocked by the farmer to harvest early crop. Normally magur is extensively cultured in West Bengal in conjunction with Indian Major Carps @ 7500-10,000 numbers/hectare.

## Breeding Method and Hatching Technique

Despite large-scale natural collection of magur, there is a dearth of quality seed for intensive/semi intensive culture. Artificial propagation of magur was tried with success at Mogra (West Bengal) by digging of horizontal holes (20-30 cm) inside the tanks below the water surface. The holes are filled with straws as substrate for adhesive eggs. One pair (male and female) of magur for each hole was liberated for breeding. No special care was taken for hatchlings. During the months of September to November, magur fingerlings were retrieved from tanks. The procedure as in vogue with little modification in Thailand was followed.

## Artificial Propogation

Magur seed is produced through hypophysation (carp pituitary extract at the rate of 30 mg/kg) and also by hormonal as substitute to hypophysation.

## Fertilization

Pairs of male and female fish were administered with ovaprim. Ovaprim was used at a dose of 2.0 to 2.5 ml per kg body weight of female and @ 1ml per kilogram body weight of male. The female ones attained the peak stage of ovulation after 15-16 hours of ovaprim administration when they were stripped off the ova in a tray. Simultaneously the grinded testis containing the milt of the injected males was thoroughly mixed with the stripped ova to obtain fertilization.

## REFERENCES

Das, Debabrat, P. Sahu and A.K. Reddy, 1999. Indian and Thai magur culture: Present status, perspectives and constraints. *Fishing Chimes*, 19(8): 11–15.

Francis, T., 1997. Hormones and spawning in catfishes: A review. *Fishing Chimes*, 17(8): 21.

Lage, A.T., Vekateswara, K. and Venugopal, G., 1993. Field trials to culture of *Clarias batrachus*. *J. Indian Fish. Asso.*, 23: 15.

Thakur, N.K. and P. Das, 1986. Synopsis of biological data on magur, *Clarias batrachus* (Linnaeus, 1958). Bulletin No. 41, April 1986, Central Inland Fisheries Research Institute, Barrackpore, West Bengal.

*Chapter 15*

# White Spot Disease (WSD) in Prawn

Prawns having very high percentage of protein. It provides several tones of high-class animal protein for consumption within country annually. It forms the major items of export of Indian marine products. Prawn fishery is definitely playing an important role in the Indian economy. It is due to its increasing demand in the foreign markets. Our country is a 'world leader' in the prawn production and to exporting along with Japan, USA, Philippiness, South American countries etc. Prawn fishery in India has already taken the shape of a regular industry. According to Marine Products Export Development Authority (1999) among the export at marine products, cultured prawns contributed 52 per cent in terms of quantity and 75 per cent in terms of value of the total export earnings from prawns.

During the last three or three and half decades there has been considerable advance in the developed countries in the knowledge of diseases affecting the prawns. White Spot Disease (WSD) has been reported in a variety of crustaceans, including wild and cultured penaeid prawns, crabs and cray fish. WSD first appeared in Northeast Asia in 1992-93 and rapidly spread to several prawn farming countries in Asia and the Indo-Pacific. Natural infections of WSD have been observed in several commercial penaeid species including *Penaeus monodon, P. japonicus, P.chinensis, P. indicus,*

*P.merguiensis* and *P.setiferus*. Since its appearance in Asia, WSD has caused major losses to the prawn culture industries in several countries including India, China, Japan, Thailand, Indonesia, Sri Lanka, Malaysia, Bangladesh and others. China alone reported losses of US $ one billion in 1993 due to WSD, and Thailand also reported loss of $ 500 million dollars from the same disease. During year 1995 this disease becomes very much rampant in brakishwater sector of West Bengal. The first outbreak of WSD in prawn along the Karnataka coast was noticed during April 1995.While its occurrence in state of Goa noticed in farms where the prawn farmers had stocked prawn seed purchased from Tamil Nadu during March 1995 and also during February 1996 from Kerala hatchery. The occurrence of disease was as a result of the infected seed brought from Tamil Nadu and Kerala region. The disease was also reported from the states like Andhra Pradesh, Karnataka, Orissa and Maharashtra.The disease brought about the healthy toll in prawn production. The intensity of the infection was so grave that it did not spare the large, medium or small-scale prawn culture systems. Experiments carried out in laboratory and also in fish farms and traditional bheries at the different tracts of North and South 24-Paraganas (West Bengal) clearly indicates that the alkaline aquatic environment and high salinity augments the bacterial activity responsible for several prawn diseases. The prawns were found externally ruptured and diffused alimentary canal and white spot on carapace. During the study conducted by Department of Fisheries, West Bengal government, a total of 150 samples were examined bacteriologically. All the organs of 88 prawns yielded a heavy growth of vibrio parahaemolyticus, which was confirmed by double ring test. A pathological change shows tissue invasion and necrosis suggesting active invasive disease of the prawns by vibrio parahaemolyticus strains. The disease occurs through excessive use of antibiotics, pesticides and feed with a very high nitrate content which may cause the alkalinity of water to rise. The bacteria and fungus also get more active. It is also reported that the WSD is primarily a bacterial disease but often-viral infection are associated with it. Fungal infection also appears secondarily. The disease normally spreads after winter with the Celsius scale around 26 degrees when the phosphate level is low. The overstocking of prawns is often one of the causes of disease as it leads to oxygen deficiency. The infection grows rapidly and contaminates other animals. The prawns infected with WSD have ruptured and diffused alimentary canal and white spots on them. The rotten prawns have foul smell and enteritis. Prawns can be consumed in the initial stage of the infection, as cooking kills the all bacteria. Besides, the prawn molts and sometimes get rid of the disease.

Experiments were conducted by Jankiram *et al.* (1996) to study the localization of the infective agents in the various tissues of the *Penaeus monodon*. They revealed the pronounced histopathological changes in the carapace, gills, muscle and hepatopancreas. The WSD affected prawns carried, apart from virus infection, heavy infections with ciliated protozoans and filamentous bacteria on the gills and appendages and *Vibrio sp*.in the hepatopancreas indicating it to be a syndrome.

Raman *et al.* (1996) also made observations on histopathological changes associated with WSD in *P.monodon* from shrimp farms of coastal Andhra Pradesh.They found hepatopancreas as a principle target organ of infection. The main visible sign of the disease after characteristics 'white spot' was the pale yellowish liquefaction of hepatopancreas. Massive necrosis of the hepatopancreas tubule, epithelial cells and intramuscular inclusion bodies in the hepatopancreatocytes were observed. Opacity of abdominal musculature and multifocal necrosis of muscles and gill epithelium and cuticular epidermis were noticed. Detachment of epithelial cells of hepatopacreatic tubule, midgut and hindgut were observed by them.

Carlos Lara (from Costa Rica) put together a package of information gathered on WSD during his visit to Prawn farming areas in Thailand. According to Lara the main disease vector is wild seed. Infected, laboratory seed is also an important vector, and it is important that laboratories undertake a monitoring and certification programme for its seed. It is not enough to use disease free seed; it is also critical to eliminate other crustaceans that are carriers of the viruses. In Thailand the recommended procedure involves completely filling up the production pond and then using a fast-acting, short-lived insecticide. The rationale is to quickly eliminate all crustaceans in the pond. It is believed that the White Spot Syndrome Virus (WSSV) does not live freely out of a host for more than 3 to 4 days, there is a high degree of certainty that five days after application of the insecticide the pond is free of viruses and can be stocked with disease free seed. In Thailand there are reports of WSSV outbreaks following a rapid change in pH or salinity. It is also recommended to reduce or eliminate pond stocking during cold months, because apparently the incidence of WSSV is higher at low temperature.

According to Mishra *et al.* (1996) the histopathological analysis of hepatopancreas and gills showed necrosis and degeneration of tubules and presence of eosinophilic bodies indicating viral infection. Primary infection with virus and secondary bacterial infection was thought to be the cause of large-scale mortalities.

## Symptoms of WSD

1. Just before occurrence of WSD, the prawns showed abnormally high growth rate.
2. Prawns swim in-group and aggregate along dykes.
3. Prawns sometimes jump out of the water.
4. Firstly *P.indicus* is affected, later on infection occurs in *P.monodon.*
5. Prawns cease to take supplementary feed but become more inclined towards taking natural feed.
6. Before the onset of the WSD, firstly the signs of zoothamnium infestations are observed at the base of rostrum, which rapidly spreads over the carapace, and then the WSD appear.

## Preventive Measures

1. Stocking of healthy post-larvae from noninfected environment.
2. Before culture, ponds should be sun dried and the bottom muck should be scrapped.
3. Proper stocking density should be maintained. Overcrowding should be avoided. Lower density stocking reduces the stress on shrimp.
4. Overfeeding should be avoided.
5. Entry of wild prawns, fishes and other aquatic animals should be checked.
6. Before releasing water into nature, it should first be passed into a biopond for treatment.

7. Water from an infected farm should not be released into the natural waters without disinfecting.
8. Prawns should not be fed with trash fish or other fishery by-products.
9. Sick animals should not be released into the environment. Dead prawns should be disposed off properly.

## Treatment of WSD

1. Disease can be controlled by applying Neem oil cake @ 4.5 kg/biga/mt water depth or 3.5 ppm/week. Neem has a medicinal value. In India it is mostly used for medicinal purpose and for pest control. Neem oil cake extract exhibits ability to control various fungi which may sometime produce secondarily infection to affected prawns.
2. Ascorbic acid protects collagen from reactive oxygen species induced oxidative degradation. Ascorbic acid prevents reactive oxygen species mediated microsomal lipid per oxidation and protein degradation. Vitamin C is probably a potential savior against degradative diseases. Ascorbic acid (Vitamin C, fishery grade)@ 1 gm/kg feed for 15 days is found effective for control of WSD in prawns.
3. Methlene blue is found to be effective for controlling WSD.This drug was found to have pronounced activity at 8-100 mg/ml levels. Methylene blue @ 12 gm/biga/mt water depth or.009/7-15 days interval.
4. An experiment was conducted by Raman (1996) for control of WSD in *P.monodon* by using plant latex of *Calotropis gigantea*. Infected *P.monodon* juveniles were kept in two separate group of ten each in triplicate. One set was treated with 0.2 ppm fresh latex (milky juice) derived from plant as weekly for consecutive five weeks, while the other set was kept as control. There was no mortality in the treated species while there was 80 per cent to 100 per cent mortality in the untreated ones. The treated specimens showed significant recovery. In most cases white spots disappeared or diminished significantly after subsequent moulting. Their growth also improved considerably with increasing weight.

5. 'SLCURINIUM'-a patent liquid was prepared by Chondar (1996) out of human urine.

The use of this medicine resulted in completed control and cured WSD in *P.monodon* and *P.indicus*.The drug can be applied by soaking in supplementary diet and its rational feeding four times a day stopped complete mortality of the infected prawns within 48 hours and cured the disease, over a month period of time depending upon the stages and density of infection. He further mentioned that the best results could be obtained when the medicine is used as preventive measure by feeding 'SLCURNIUM'-soaked diet to the culture stock right from their larval stages as a health tonic when the animals had grown with great vitality and became self-immuned from attack of any disease. It is also recommended for control of diseases in fish and prawns. However the proper maintainence of water quality of the culture pond is also an essential requirement as a part of this therapy to avoid stress factors and prevent further emergence of disease in the media.

## REFERENCES

Chondar, S.L. and P. Srinivasa Rao, 1996. 'SLCURINUM' to control and cure white spot viral disease in penaeid prawn. Abstract book of *National Seminar on Diseases in Aquaculture* (5-6 March), at Aquaculture Research and Training Center of CIFE, Kakinada (Andhra Pradesh), p. 13.

Janakiram, P., R. Madhavi and L. Jayasree, 1996. White spot disease of *P.monodon*: Histopathological and experimental studies. Abstract book of *National Seminar on Diseases in Aquaculture* (5–6 March), at Aquaculture Research and Training Center of CIFE, Kakinada (Andhra Pradesh), p. 10.

Mishra, S.S., M.S. Shekhar and B.P. Gupta, 1996. Incidence of white spot viral infection in shrimps in coastal Orissa. Abstract book of *National Seminar on Diseases in Aquaculture* (5-6 March), at Aquaculture Research and Training Center of CIFE, Kakinada (Andhra Pradesh), p. 8.

Raman, R.P., 1996. Control of white spot disease in *P.monodon* by plant latex of *Calotrpis gigantea*. Abstract book of *National Seminar on Diseases in Aquaculture* (5-6 March), at Aquaculture Research and Training Center of CIFE, Kakinada (Andhra Pradesh), p. 14.

Raman, R.P., G. Rajeshwari, P. Srinivasa Rao and V.N. Acharyulu, 1996. Observations on histopathological changes associated with white spot disease in *P.monodoin fabricus* in shrimp farm of coastal Andhra Pradesh. Abstract book of *National Seminar on Diseases in Aquaculture* (5-6 March), at Aquaculture Research and Training Center of CIFE, Kakinada (Andhra Pradesh), p. 11.

*Chapter 16*

# Epizootic Ulcerative Syndrome (EUS) in Fishes

Like other animals, fishes are also subject to many diseases. The disease outbreak is generally due to high stocking density, overfeeding and manuring. Fish disease in manifested by the appearance of structural or behavioral abnormalities. The parasitic organisms attack fishes of all ages. The aim of the present chapter is to highlight the Epizootic Ulcerative Syndrome (EUS), a disease responsible for massive fish mortality resulting in financial loss to the concerned. The alarming rate of fish mortality had robbed fishermen from their daily bread and caused repulsion to fish consumers from taking fish. The disease was firstly reported in year 1972 from Australia. It had quickly spread out mainly amongst mullets, murrels, perches, catfishes and carps in several south east Asian countries including India and become an alarming concern to scientists, fish farmers and consumers. The disease was caused serious damage to India's aquaculture. It appeared for the first time in India in year 1988 and has covered almost the entire country. The occurrence of EUS has been reported from various states like Maharashtra, Andhra Pradesh, Uttar Pradesh, Madhya Pradesh, Bihar, West Bengal, Orissa, Karnataka, Tamil Nadu, Rajasthan, Haryana, Kerala, Assam, Manipur, Tripura, Meghalaya, Mizoram and Arunachal Pradesh.The disease was recorded from culture as well as capture sector, which includes the rivers, reservoirs, lakes, culture ponds, irrigation canals, eels as well as the estuaries.

## Affected Fishes

### Freshwater Fishes

*Cyprinus carpio, Hypopthalmicthys molitrix, Ctenopharyngodon idella, Aristichthyes nobilis, Catla catla, Labeo rohita, Cirrhinus mrigala, Rhinomugil corsula, Amphipnous cuchia, Channa sp, Mystus spp, Heteropnestus fossilis, Clariaus batrachus, Ambassis ranga, Puntius spp, Glossogobius giuris, Anabus, Mastacenbelus sp, Notopterus sp, Rasbora sp, Amblypharyngodon mola, Trichogaster sp, Gadusia chapra.*

### Brakishwater Fishes

*Mugil spp, Etroplus sp, Platycephalus sp, Epinephelus sp, Scatophagus sp.*

#### Symptoms

Abnormal swimming behaviour with head projected out of water and floating of fish listless near the bank are the major symptoms of EUS.Clinical signs and gross pathology in the affected fishes are similar in almost all the species with moderate to severe ulcerative skin lesions. Initially the infection commences in the form of multiple inflammatory re spots on the body causing localized haemorrhage. In carps these appear within the scale pockets. In advanced stages of infection, the ulceration covers large areas with sloughing of scales and degeneration of epidermal tissue. With further advancement of the disease, ulcers become deep haemorrhogic and necrotic often with black melanistic rim. In advanced stage, large and deep ulcers are commonly seen in all parts of fish, especially on head, abdomen and peduncle.

Generally the air breathing fishes are the first to be affected followed by weed fish, minor carps, Indian major carps and exotic carps. By preventing the air breathing fish from entering into the ponds from near by river field keeps the EUS away.

## Causative Agents

The suspected causative agents are the bacteria, virus and fungus. It is widely suspected that a biological infectious agent is the primary cause of EUS and certain abiotic factors are also responsible for creating the stress to fish. The research work carried out in India on different probable causative agent for EUS revealed that the water quality and anthropogenic factors like pesticides, fertilizers and heavy metals are also responsible for outbreak of disease. The studies conducted all over the world confirmed that the low alkalinity, hardeness, chloride concentration and pH fluctuation showed a close link with EUS outbreak. Generally the outbreaks of EUS in India occurred after the rainfall. This observation is in agreement with reports from other foreign countries leading to suspicion that drainages of agricultural chemicals may have an important role as predisposing factor for EUS outbreak. The extent of pollution may create a stressed condition for aquatic life and may be the predisposing factor for EUS outbreak. Along with some viruses, the bacteria like *Micrococus, Citrobacter, Acinebacter sp, Streptococcus sp, Shigella sp, Arthobacter sp, Pseudmonas, Staphylococuss* and *Baciillus sp* are also closely associated with the outbreak of EUS. Several fungal species like *Saprolegina sp* and *Aspergilus sp* were also isolated from the lesions of EUS affected fishes. Among animal parasites ciliates, copepods and trematodes were encountered from EUS affected fishes.

## Remedial Measures for EUS

The primary causative agent of EUS is not known exactly. It is a main difficulty encountered in counting the disease outbreak. As the EUS was observed in rivers, estuaries, tanks and reservoirs, it was not possible to treat the infected fishes. The following remedial measures are effective for curing EUS outbreak:

1. Central Institute of Freshwater Aquaculture (CIFA), Bhubaneshwar (Orissa) has brought out a product 'CIFAX' to control the disease. The small sized tanks or ponds can be treated with this drug. The treatment was reported to have caused marked improvement in the ulcerative conditions of the fishes in the small sized water bodies.

   The drug applied at 1 liter/hectare meter of the water area. The EUS in the pond is reported to cure affected fishes within 7 days.

2. Some fishermen use mostly the herbal mixtures of turmeric with neem and lime to control EUS. However there is no systematic observations on the impact of this treatment.

3. Sankararista: A herbal drug was developed by Chondar (1996). It is a liquid herbal medicine prepared by Kaviraj

Eeswar Chandra Choudhary and sold in the market in this trade name-cures the piscine EUS completely without any adverse effect to other biota of the culture media. The drug when applied mixing directly in the pond water @ 0.3 to 0.4 ppm in three consecutive doses at an interval of 2-3 weeks, depending upon the density of infection, with subsequent two 6 monthly follow-up dose can fully cure the EUS showing no further attack of the disease in the same water.

4. Bleaching powder at 1 mg/liter or 5-10 kg/hectare was reported to be useful in cure of initial lesion of EUS.
5. Application of Nacl at a concentration of 3-4 per cent dip treatment of affected fishes have given fairly effective results for curing early stage of the disease.
6. Quick lime @ 100-1000 kg/hectare has been found effective for curing EUS. Lime is used according to the pH of water.
7. Application of $KmnO_4$ applied at the rate ranging from 1ppm to 10ppm.It gives fairly good results. While the application rate for bath treatment of fish is 1 to 6 ppm is recommended.
8. A microencapsulated feed containing 30 per cent protein, Nalidixic acid, Erythromycin along with vitamin A and C has been formulated by Central Inland Fisheries Research Institute (CIFRI), Barrackpore (West Bengal), use of this pelleted feed to diseased fishes showed the fishes recovering. It was found that the antibiotics either erythromycin or oxytetracycline or tertramycin at 60-100 mg/kg of feed for 7 days can cure the ulcers of EUS affected fishes.

## REFERENCES

Chondar, S.L., 1996. 'SANKARARISTA' a herbal drug ensuring cure of EUS disease. Abstract book of *National Seminar on Diseases in Aquaculture* (5-6 March), at Aquaculture Research and Training Center of CIFE, Kakinada (Andhra Pradesh), p. 31.

Das, Manas K., 1997. *Epizootic Ulcerative Syndrome in Fishes: Its Present Status in India*. Central Inland Fisheries Research Institute, Barrackpore, Bulletin No. 69.

*Chapter 17*

# Hilsa: The King of Fishes

## Local Names

Hilsa (Hindi); Oolul(Tamil); Polasah(Telgu); Palla(Sindhi); Ilish(Oriya and Bengali); Ilihi (Assamese); Paliyah (Malayalam and Kannada).

## Introduction

In ancient Sanskrit writing hilsa was glorified as the king of fishes and its taste is said to surpass nector. It is also known as 'Indian shad'. In Bengal it is choicest among fish delicacies and a pair of hilsa is offered to Goddess Durga on Vijaya Dashmi Day.Hilsa enjoys a very wide distribution in marine and fresh water gradient all along the Indian coasts. The migratory habits of hilsa make it a unique biological material capable of withstanding wide range of salinity and covering a long distance against strong water currents. It is essentially marine fish found at a depth of about 36 meters in areas ranging from Persian and Arabian Gulfs, Eastern Africa, Indian subcontinent, East China, Malaysia and Thailand. In India Hilsa is distributed in the foreshore and offshore areas of east and west coast, estuaries and rivers of Ganga, Brahmaputra and Hooghly-Matlah estuary, Mahanadi and Chilka lagoon, Godavari and Krishna, Pennar on the east coast and the Narbada and Tapti on the west coast.

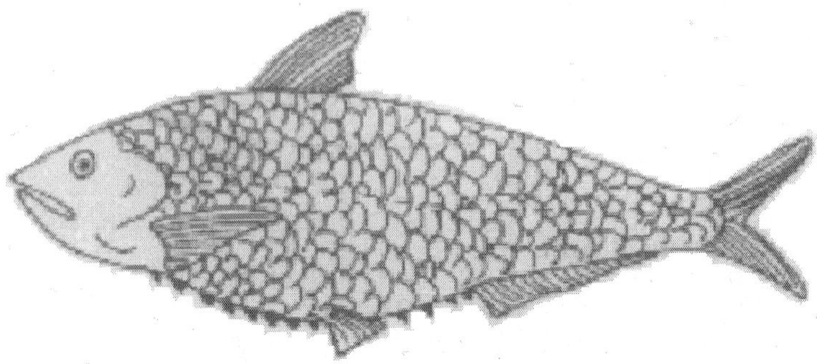

**Figure 17.1: *Hilsa ilisha***

## External Characters

Body compressed, oblong, abdomen keeled and serrated. Head large, mouth terminal, maxilla extends up to the post-orbital margin. Eyes large, lateral with adipose lid, in anterior part of head and not visible from below ventral surface. Lips moderately thick, jaws subequal with a distinct median notch at the symphysis of the upper jaw. Dorsal fin originates in advance to the place of origin of ventral fin. Anal fin single with moderate length. Caudal deeply forked. Scales large, lateral line system absent. Barbels absent, colour silvery white with transverse faints black bands along the dorsal side extending up to the middle of the body.

## Food and Feeding Habits

Hilsa is mainly a plankton feeder. Hilsa fry (20-40 mm) mainly feed on the diatoms, copepods, daphnia and ostracods whereas the young Hilsa (up to 100 mm) feeds on smaller crustaceans, insects and polyzoa. Hilsa feeds at the bottom. The gills and gill rackers are modified into very fine sieve that must filter all microscopic food organisms entering the buccalcavity with respiratory water current.

## Age and Growth

Hilsa grows fairly fast in early stages and attains a length of 35 mm in one month and 130 mm in four months. At the end of fifth month a size of 160 mm is attained and thereafter the fish grows at the rate of about 20 mm per month. The fish reaches a maximum size of about 300 mm in one year.

## Maturity and Spawning

Hilsa attains sexual maturity at the age of one or one and half years. While the investigations by De (1986) reveals that the maturity of females occurs at about 341 mm length and the maximum values of Gonadosomatic Index (GSI) were recorded during September to March with a peak in October, indicating its gondal maturity and breeding season. The more recent studies (De, 1980) revealed that the presence of only a single size group of ova in mature ovary indicating a definite spawning season of fish. Fertilization is external. Eggs are demersal, hatching takes place in about 18 hours.

## Fecundity

It lays 2, 50,000 to 1, 600,000 eggs depending upon the size of female fish.

## Population

A comparative study by several workers confirmed that hilsa populations in the river Hooghly and other rivers differ significantly from one another. A large sized fishes mainly contribute the monsoon hilsa fishery, while during winter period small sized groups forms the main fishery.

## Migration and Fishery

Hilsa travel long distances in the sea and run up to the rivers. Such type of migration is known as 'anadromous migration'. The dams and barrages form physical obstruction to the migratory fishes causing physiological strain and breeding failure. The classic example is hilsa, which has a migratory range of 1500 km upstream of the estuary. Construction of Farakka barrage in West Bengal, 476 km from the river mouth has nearly eliminated the hilsa fishery above the barrage. The impact of Farakka barrage on hilsa had been very severe. The average landings of hilsa during pre and post Farakka period were studied in detailed by Chandra (1989):

| *Centre* | *Pre-Farakka* | *Post-Farakka* | *% Decrease* |
|---|---|---|---|
| Allahabad | 19.30t | 1.04 t | 94.61 |
| Buxar | 31.97 t | 0.60 t | 98.12 |
| Bhagalpur | 03.95t | 0.68 t | 83.05 |

The Hilsa fishery was declined by 94.61, 98.12 and 83.05 per cent at Allahabad, Buxar and Bhagalpur respectively. Thus, the construction of Farakka barrage has greatly affected Hilsa migration leading to the collapse of its fishery in the middle stretches of river Ganga. Attempts are being made to artificially propagate this fish with a view to fostering recovery of the lost fishery. In Pakistan the Ghulam Mohammed barrage constructed in year 1954 on the Indus near Hyderabad 291 km upstream from the river mouth, deprived hilsa at approximately two thirds of their previous spawning area.

Hilsa fishery is an important fishery of Chilka lagoon, Hooghly-Matla estuary as well as the lower reaches of most major rivers. In Chilka lagoon it forms a regular fishery throughout the year, it is mostly confined to the post monsoon period in the estuaries of Mahanadi and Ganga when breeders migrate up the stream in large number. However, Hooghly, Padma, Ganga and Godavari on support separate monsoon and winter fisheries and average a good hilsa fishery yields about 16000 metric tons in a successful year. The catch during one single operation may run into hundreds of tons particularly when the fish is moving in shoals.

Hilsa fisheries on eastern bank of river Godavari provide employment to about 1500 fishermen families from Yanam to Dowlaiswaram in the month of July to October. The average landing on the bank of the river ranges between two to three hundreds tones per season. However, it is highly fluctuating and depends upon monsoons, siltation of the river that reduces its depth. Hilsa catches are dwindling year by year.Effective measures have to be taken to restore hilsa. It is felt that Hilsa in freshwater is more tasty when compared with estuarine hilsa. Hilsa have very high demand among Kolkata fish consumers. Howrah wholesale fish market is the largest fish market in India and the second largest in Asia. A case study of the same fish market revealed that the hilsa fetch very fancy price particularly during the Vijayadashmi day and marriage season, but unfortunately the supply of this fish is very meager. The fishes are sold in three size groups *i.e.*, small, medium and large.

## Remarks

The depletion of hilsa fishery in many river stretches of the country has been observed due to the construction of dams, weirs and barrages across the rivers which act as barriers in the migration of the fish. Therefrore, there is an urgent need for conservation as well as protection of natural breeding grounds of the fish. Further Central Inland Fisheries Research Institute (CIFRI), Barrackpore (West Bengal), which would certainly assist in enhancing production from the rivers and estuaries by implementing certain management measures, should carry out the investigations on biology of fish.

## REFERENCES

Chandra, Ravish, 1989. Riverine fisheries resources of the Ganga and the Brahmaputra. In: *Conservation and Management of Inland Capture Fisheries Resources of India,* (Eds.) Arun G. Jhingran and V.V. Sugunan. Inland Fisheries Society of India, pp. 52–60.

De, D.K., 1980. Maturity, fecundity and spawning of post-monsoon run of *Hilsa ilisha* (Hamilton) in the upper stretches of the Hooghly estuarine system. *J. Inland Fish. Soc., India,* 12(1): 54–63.

De, D.K., Amitabha Ghosh and V.K. Unnithan, 1989. Biology and migration of Hooghly hilsa in the context of Farakka barrage. In: *Conservation and Management of Inland Capture Fisheries Resources of India,* (Eds.) Arun G. Jhingran and V.V. Sugunan. Inland Fisheries Society of India, pp. 197–201.

*Chapter 18*

# Fisheries Cooperatives in India

The cooperatives in India have made remarkable progress in the various segments of Indian economy. There are 5.03 lakh cooperative societies with a membership of more than 20.9 crores and working capital of more than Rs.227111.8 crores. In many segment of Indian economy, cooperatives are contributing predominant role in the credit sector despite of keen competition from nationalized and private sector banks. Cooperatives are disbursing nearly 46 per cent of total agricultural credit. They are distributing 36 per cent of total fertilizers in the country. The share of cooperates in sugar production is nearly 55 per cent , in spindlege 10 per cent , in yarn production 22.1 per cent and in coverage of handlooms 55 per cent. The cooperatives are playing a very important role in the public distribution system to serve the weaker sections of the community. Nearly 28 per cent of the rural fair price shops are within the cooperative fold. Cooperatives have also played an effective role in exports. The economic reforms, which have been introduced since 1991 have given new dimensions to precepts and practices of economic development. However, cooperatives have not been able to take the fullest advantage of the economic reforms, as they have not been provided level playing field.

The cooperative sector has always remained a center of attraction in India. The cooperative movement has been perceived in many ways and it has many meanings to many people. For an economist, it is an economic activity, for a sociologist, it is a change agent, for a social worker, it is a social work and for a politician it is a center of power.

In the early stages, the cooperative movement was perceived as a welfare activity. But after independence it was accorded state sponsorship and the voluntary nature of the cooperation got increasingly minimized over a period of time and the system was increasingly politicised.

Cooperatives are the shield of the weak. Fishermen are among the weakest sections of the community. Illiteracy, poverty, and lack of knowledge of latest fisheries technology are contributing factors. This vicious circle is further strengthened by lack of institutional support, both in infrastructure and finances. Consequently, middlemen, who act as moneylenders, traders and contractors, subject fishermen to exploitation.

Fishermen discovered cooperatives could spare them from exploitation and improve their socio-economic conditions. Efforts made in this direction have yielded good results in some areas, but the overall picture of fishery cooperatives is not encouraging.

## Fishery Cooperative Movement

The Indian fishery cooperative movement began in year 1931 when the first fishermen's cooperative society was organized under the name of 'Karla Machhimar Cooperative Society' in Maharashtra. State West Bengal was the next to organize cooperative societies in fishery sector in year 1918. In the same year Tamil Nadu also organized one cooperative society. The structure continued to grow over years into multifunctional units at the primary level, federations at district/regional, state and national levels.

A study conducted by the council for social development confirmed the suitability of the fishery cooperatives, "as a tool for promoting the interests of fishermen". The study also emphasized the need for, " organizing active fishermen into cooperatives and for strengthening and encouraging the fishermen's cooperative societies for performance of multi-purpose functions and social interest of their members". To ensure that a well-knit structure of fishery cooperatives is created in the country, the study recommended that, "multi-functional primary cooperatives can be retained".

A number of fishery cooperatives in the country are helping their members and their family members to the extent of providing complete marketing infrastructure for the sale of the catch at remunerative prices.

In some of the states, the fishery cooperative movement is working very effectively and a number of evaluations have confirmed the efficiency of these organization.One evaluations confirmed that arrangements of marketing made by the fishery cooperatives in the state of Maharashtra save the members from exploitation.

## Problems and Solutions in the Fishery Cooperatives

1. The cooperative fisheries sector in India faces crucial problems. Existing cooperative law does not support fisheries cooperative development. Countries like Japan and South Korea have special enactments, which guarantee allotment of fishing waters to fisheries cooperatives and ensure membership of genuine active fishermen. The law defines the roles within fisheries cooperatives.
2. In most states special provisions in the cooperative societies act or a separate act for cooperative land development banks (now known as agricultural and rural development banks) exist. In order to promote fisheries, the state governments should formulate separate provisions for fisheries cooperatives within the act or enact separate legislation to allot water bodies and avoid overlapping operation, finances, and structural linkages among fisheries cooperatives in the state.
3. Those states, which have not yet organized a federation of fisheries cooperatives, should begin one with sufficient equity to take up business and promotional activities for fisheries cooperatives in the state. Existing state level federations must be activated.
4. At the regional level, the gap is very wide in these states. In light of the establishment of fish farmer development agencies inland and brakishwater fisheries, it is necessary

to organize regional/district level fisheries cooperative federations to provide inputs, operational inputs, harvesting and marketing support to the members of primary fisheries cooperatives and fish farmers.

5. At the primary level in most state fisheries cooperatives overlap. This results in unhealthy competition between them, particularly for water bodies. State governments should take necessary steps to correct this, and to organize/ reorganize the existing primary level fishery cooperatives.
6. It should also issue clear-out instructions to allot water bodies to re organized cooperatives so that they may have necessary fishing water to provide employment to their members.
7. In the marine sector, the primary fisheries cooperatives should be strengthened with infrastructure facilities like landing centers, market yards, roads, transport facilities etc. Fisheries cooperatives should be given necessary support to pull their catch and have access to strong export infrastructure with qualified staff.
8. Women also play an important role in fisheries cooperatives. Once the fishermen bring the catch, their job begins. Cooperative law does not allow membership of both husband and wife, which prevents women from participating in the management of the fisheries cooperatives. Women involved in the fishery must participate in the management of fisheries cooperatives.To market fish profitably, ensuring a fair price for both producers and consumers; cooperative marketing in India must be developed. Funding for infrastructure should be provided by central, state and local bodies, and then handed over for management to fishery cooperatives.
9. Cooperative banks and other financial institutions have not helped to develop fisheries cooperatives in most of the states. Financial aid is often not available where it is needed, like payment of lease money, welfare and credit. NABARD should also take immediate steps to encourage cooperative banks to provide credit to fisheries cooperatives where it will be most effective.

10. The efficiency of cooperatives in respect of both organizational and financial aspects through professional management may be enhanced through appropriate steps. The National Federation of Fishermen's Cooperatives Limited (FISHCOPFED) may give necessary training to office bearers of the cooperatives. The federation may be supported by the Department of Agriculture and Cooperation (DAC)/Department of Animal Husbandry and Dairying (DAHD) in the Ministry of Agriculture (MOA) for having a panel of experts with suitable remuneration and other allowances.
11. FISHCOPFED may ensure close functional linkages between the cooperatives at village, district and apex levels through year round interactive programmes appropriately incorporated in their plan documents. Government of India may consider enlarging share capital base and budgetary provision to meet the challenges of additional activities.
12. Fishermen's cooperatives and Self Helping Groups (SHG) need to be encouraged. Incentive schemes need to be introduced by the center/state governments to promote capture fisheries and aquaculture in cooperative sector through low-cost technologies and training impaired by the FISHCOPFED with the help of DAC/DAHD in the MOA.
13. Overcapitalization being one of the adverse outcomes of liberal credit provision particularly in case of marine fishing fleet development, it is necessary to reduce debt burden by adopting appropriate technologies and control over operational expenses by the respective fishermen's cooperatives.
14. Unlike agriculture, there is no support to rescue fisher folk and fish farmers during times of natural calamities, which cause damage to their catches, infrastructures and equipments. The governments may extend protective provisions as in agriculture sector to capture fisheries and aquaculture activities.
15. An exercise may be initiated by FISHCOPFED to determine and standardize unit size of infrastructures and resources

to ensure techno economic viability of fishermen's cooperatives.For this purpose; a team may be constituted for concluding a techno-economic study.

16. Exemption of excise duty on HSD oil used by fishing boats (of less than 20 mtrs length) in marine sector as at present may also be extended to inland sector. In order to avoid misuse, the scheme should be made applicable in favour of fishermen's cooperatives only with essential checks and controls.

## Schemes of NCDC in the Development of Fisheries

NCDC has been promoting and developing fisheries cooperatives after its Act was amended in 1974 to cover fisheries within its purview. The Corporation has formulated specific schemes and pattern of assistance for enabling the fishery cooperatives to take up activities relating to production, processing, storage, marketing, etc. Assistance is provided to fisheries cooperatives on liberal terms treating the activity as weaker section programme. Assistance to fishery cooperatives is provided for the following purposes:

## NCDC's Assistance

Till 31-03-2004 the Corporation has sanctioned assistance of Rs.86482.06 lakhs and released Rs.59390.71 lakhs for fisheries development through cooperatives in different States and UTs.

## Dovetailing of MFPI's Assistance with NCDC's Assistance

NCDC has dovetailed the Scheme of the Ministry of Food Processing Industries for preservation and processing of fish with its assistance and has obtained grant-in-aid approval of Rs.1319.801 lakhs for projects in the states of Maharashtra, Karnataka, Tamil Nadu, West Bengal, and Kerala. The assistance under the scheme has been made available for insulated boxes for retail fish marketing, transport vehicles (mopeds, auto-rickshaws, mini trucks, etc.), fish drying rafts, retail outlets, processing units, ice plants, cold storages, etc.

## Dovetailing of DAHD's Assistance with NCDC's Assistance

NCDC has dovetailed the Schemes (for Motorisation of traditional fishing crafts, introduction of intermediate deep sea fishing vessels, safety at sea, infrastructure facilities for preservation of fish, fish landing centres and fishing harbours) of the Department of Animal Husbandry and Dairying, Ministry of Agriculture with its assistance for cooperatives for the 10 th five year plan.

## Conditions of Eligibility for NCDC's Assistance

1. The society seeking financial assistance should be functional, working efficiently.
2. A newly registered society can also be considered, provided it has a well designed programme of pisciculture, marketing etc.
3. Primary societies should be affiliated to the regional/state level fisheries cooperative federations and the district/ state cooperative union, if existing in the state.
4. The society which has been classified below "C" category during the audit for the last three years will normally not be considered. However rehabilitation projects of such societies, those would be considered on merits.
5. Societies with negative net worth (cumulative losses exceeding own funds) would not be eligible for assistance.
6. The inland fishery societies operating in the water area leased out to them should invariably have a lease for more than three years. For availing assistance for excavation , the period of the lease should be for a minimum of 10 years.

## Impact of NCDC's Assistance

1. The increase in marine fish catch of cooperatives in Gujarat, Kerala, Maharashtra, West Bengal, etc., is attributed to the support of NCDC.
2. The total number of fishermen to be benefited from all the Integrated Projects is estimated to be over 2,84,100.

3. The total estimated additional fish production on implementation of the projects is over 3.52 lakhs tonnes per year.
4. The societies are democratically managed and the members take active interest in the functioning of the societies.
5. Beneficiary fishermen earn more income than non-beneficiaries. NCDC's assistance has helped fishermen by reducing drudgery of work, increase in the number of fishing days thereby increase in fish catch of fishermen. Fish marketing infrastructure helps the fishermen's cooperatives in realising better value for fish catch in the market.
6. The increased fish catch by the fishermen has improved the socio-economic conditions of the fishermen perceptibly.

## National Bank for Agriculture and Rural Development (NABARD)

The main object of National Bank for Agriculture and Rural Development is to extend financial support to the development of marine fishing industry to enable the industry to register over a period of years, substantial increase in marine fish production. NABARD also attaches great importance to support infrastructure. Lack of facilities for handling a perishable fish production unrewarded because if share facilities do not exist the producer would not get any financial benefit of increased fish production. NABARD therefore extends financial support for creation of infrastructure to take care of fish production. Establishment of iceplants, fishmeal plants, fish auction halls, service stations and workshops, bunkering facilities and so on, are all eligible for refinance, provided the scheme includes the production oriented items, namely fishing boats. Refinance facilities are also available for transport, distribution and marketing of fish. For instance, refrigerated trucks or insulated trucks, fish stalls, and deep freezers are covered by the NABARD refinance.

*Chapter 19*

# Fish Marketing in India: A Case Study of Howrah Fish Market

Howrah wholesale fish market is the largest fish market in India and the second largest in Asia. The Tokyo Central Wholesale Fish Market, Tokyo in Japan is the largest wholesale fish market in Asia handing about 80,00,000 tonnes of fishery products annually.

Howrah wholesale fish market, Howrah is situated in the vicinity of Howrah railway station. The building in which the market is housed is an old enclosed type with its interior ramified. It is connected by two main gates with the Maulana Abul Kalam Azad Road. The market occupies about an area of 0.4 hectare. There are 180 stalls and each one is having an area of about 14 sq.m.Every stall is having a raised platform which serves as the sales counter of the wholesaler. In front of the stall there is a pucca brick and concrete flooring for keeping the fish lots at the time of auction. Transactions in the market generally start taking place at 0600 hrs and continue late in the night up to 2100 hrs.The peak period of business now –a-days is from 0800 to 1800 hrs.The market remains open for all the days of the week except Sundays. The market is visited and about 10,000 to 17,000 persons who are engaged in the retail business of fish attend auctions. The number of visitors goes as high as 20,000 during the festival and marriage periods when the fish is in huge demands.

The fish from various states starts arriving at the market right from 0400 hrs by rail and lorries. The fish is carried to the stalls of various wholesalers who also function as commission agents or auctioneers. The fish is sorted out variety-wise and graded size-wise and then put to auction by open bidding.

## Market Functionaries

The wholesalers are also the commission agents. They dispose of the fishes to the retailers or to other fish merchants whole bids are highest in the auction. The number of bidders is increased in this market at the time of marriage and festivals and on special occasions. Generally its proprietor or his representative manages the business of one stall. An account clerk known as 'munim', a cashier, and an auction clerk who auctions the fish assists him. All these members of staff are called 'gaddi staff'. Every stall has its own fish sorters and labourers who are collectively called as 'mutia'. The members of the gaddi staff are paid salary on monthly basis. The number of staff working under a wholesaler depends upon the turn over of his business. Some stall owners are receiving an ample supply of fishes from morning till evening and naturally they have got more number of staff to handle the business smoothly. The labourers or the mutias of every stall get Rs. 7 (Rupees seven) for clearance of one basket of fish.

The fish consignments are brought in two ways in the market. A person known as 'munshi' is hired by 15-25 wholesalers. He is authorized by the wholesalers to take delivery of fish on their behalf. The Munshi gets the consignments cleared from the Howrah railway station and makes necessary arrangements for its transportation to the stalls of his client wholsesalers. The Munshi also prepares the reports regarding the daily train-wise arrival of fish consignments and supplies these reports to all his client wholesalers. From these reports the wholesalers get the information regarding fish supply received by other colleagues. In this way, the wholesalers receiving fish from a single supplier try to sell it at the highest possible rate so that the sender may in future supply him more quantities in comparison to others. For unloading of the fish consignment from the train and to bring it in the market, the labour gets Rs.6.50 per basket. The labourers engaged for unloading the fish consignments from the trucks carrying it in their carts to the stalls of the wholesalers are paid Rs. 5 per basket.

Apart from above specialized labour there are others labourers doing some odd jobs. They are called as 'Jhalliwalas'. They are generally engaged by retail merchants and by other wholesale fish purchasers who send fish outside Kolkata and Howrah.The jhalliwalas take the fishes from the side of the stall to that spot where these retailers keep their purchase at one place under the supervision of another person. Jhalliwalas are paid at the rate of two or three rupees per trip.

## Preparation of Fish for Auction

The fishes received late during the preceding night are packed in fresh ice for keeping their quality and are kept for auction on the next day. Just from early morning, the fishes are unloaded from the trains and trucks and brought to the respective stalls. Major carps and catfishes are graded size-wise. There are four grades according to size; 1) small: below 1 kg; 2) medium: 1kg to 2.5 kg; 3) large: 2.5 kg to 5.0 kg, and 4) super large : 5 kg and above.

Bata, Khaira, and Bam are auctioned in a lot. Sea fishes of small sizes are also auctioned without sorting. Other quality fishes, which are graded and auctioned, are Hilsa and Bhetki.Hilsa has got two grades, one small and other large. Live fishes are received in drums. Koi (*Anabus testudineus*) is kept in separate drums and Singhi (*Heteropneustes fossilis*) and Magur (*Clarias batrachus*) are put together in the same drum. They are auctioned drum wise, but all the rates of the fishes either sale unit is 'kg'.

## Pricing of Fishes Through Open Auction System

Open auction at stalls of the wholesalers starts from six in the morning, as by this time the fish consignments reach the market. On every stall, quality fishes are sorted and graded species wise and are displayed in very attractive lots for auction.

The small fish dealers and other interested buyers attend the auction at the stalls and bid for the available fish of their choice.

The salesman (auction clerks) of the commission agents are so experienced that they guess the trend of the market on that day. If the fish supply in the market is less on that day immediately he fixes highest price in his mind at which he would accept the auction bid. The retailers and other participants start bidding and quoting their prices and thus a competition between the participants is generated. When the bidding rate reaches the preconceived rate of the salesman he lets the hammer fall by repeating thrice the last bidding rate. Now the auction is closed and that the highest bid rate becomes the market rate for that lot of fishes. The fish lot for which the auction bid has been finalized, is weighed and handed over to the highest bidder.

The retailers are given 5 per cent of rebate on gross weight of the fishes purchased by them, *i.e.*, for every 100 kg fish purchased the bidder is charged only for 95 kg.

The sale amount is collected in cash from the buyers. Some good permanent customers are sold fish on credit basis but they have to clear the amount within a week's time.

The auction prices of fishes vary from wholesaler to wholesaler during the same time and same day and it is here that the efficiency of the salesmen is counted since the suppliers prefer that every commission agent who could pay them the highest possible price of their fish. Fluctuations in the fish supply are unpredictable so there are much variation in the fish prices in the market.

## Pricing and Price Behaviour of High Demand Fishes

Fish species belonging to certain groups have very high demand in the Howrah Wholesale Fish Market. Daily prices of such high demand varieties were collected from some of the wholesalers from which the monthly average prices were computed. A brief account of the price behaviour of high demand fish varieties in the Howrah Wholesale Fish Market is given below:

## Indian Major Carps

The species of Indian major carps such as *Catla catla* (Catla), *Labeo rohita* (Rohu), *Cirrhinus mrigala* (Mrigal) are the most favoured species of the Bengalis. These species have very high demand among Kolkata consumers. The super larger sized catla and rohu fetch very fancy prices particularly during the festival and marriage seasons but unfortunately the supply of such large fishes is very meager. The price behaviour of different major carps is described below.

*Catla catla,* the fastest growing Indian major carp has mush demand in whole of West Bengal. The large sized catla are preferred during the festivals and marriages. The study of prices revealed that the fish is sold in three sized groups based on weight such small, medium and large. A study of the average monthly prices of catla reveals that the price variation in the different groups is mush more in the Howrah Wholesale Market. The largest group was sold at the highest price. The medium sized catla were sold almost at the same rate.

*Labeo rohita* is considered as one of the tastiest fish and it has a great demand in the state. The price variation in respect of small rohu was not very wide. The medium sized rohu showed variations in prices. The large sized rohu fetched higher prices.

The small sized mrigala fetched fewer prices as compared to small sized catla and rohu in all the wholesale markets. In the Howrah Wholesale Market, the prices of this group remain almost constant.

Large catfishes such as *Wallago attu, Mystus aor, Silonia silondia* and *Pangasius pangasius* are sold in the wholesale fish markets of Kolkata. They have more demand among the people belonging to the East Bengal and North India. These species are preferred as they have very less intramuscular bones and can be eaten very easily as compared to carps. Catfishes are sold in two size groups-small *i.e.,* fishes weighing below 3 kg and large *i.e.,* fish above 3 kg in weight.

Live fishes comprise three species *i.e., Anabus testudineus* (koi), *Heteropneustes fossilis* (singhi), *Clarias batrachus* (magur).All these are marketed alive and are considered to have some medicinal value. They are easily digestible, so they are prescribed in the diet of children, convalencing and old people. They showed high demand throughout the year. Koi was sold at higher prices.

The hilsa is the favoured fish of the local people and it is always in high demand. The species sold in three size groups, small, medium and large. The prices of large fishes were maximum

As assortment of various species such as *Ompak pabda, Notopterus notopterus, Gudusia chapra, Mastacembelus spp, Labeo bata* etc are considered also the important fishes in the market. Among these fishes *Ompak pabda* got the highest monthly prices and *Gudusia chapra* received the lowest prices.

## Problems in Fish Marketing

### Storage Facility

Fish starts spoilage after 7 hours of its catch. If they are not sold or stored properly within this period, it will loose its market value. All fishes are not available throughout the year. So it is not possible to supply the all fishes throughout the year in same quantity into the market. Fishes are also caught in a bulk quantity in a particular season of year. To supply fishes into the market in off-season, storage facility should be adequate. But in India, the storage facilities are not up to the mark, creating problem in fish marketing.

### Illiteracy and Poverty

Generally the fishermen are illiterate. They do not know the market position and strategies. Most of them are not mechanically sound enough with modern fishing techniques.

Fishermen are poor. For the fishing operation and avail the storage and transportation facilities, they need a large amount of money. For money, fishermen depend on the traders, and they take a large portion of share, make fisherman poorer.

### Problem of Transportation

Transport is one of the main key of successful marketing. The insulated or refrigerated vans are required to transport long distances. Sufficient transport facilities are not available in India.

### Credit Facilities

Banks refuse to provide credit to fishermen. It is because they have doubt of replacement.

### Lack of Gradation

By proper grading and standardization, buyers can make their choice. It gives greater satisfaction to customers. But grading is not available in fish and fishery products in Indian market. As we know, standardization is the popularization of grading but it is not implemented in Indian fish markets.

### Fish Handling

During catching and transportation fishes are not properly handled. About 10-15 per cent of the fishes are damaged due to mishandling.

## Market Information

There is a lack of proper market information. Fishermen do not know when, where and what is the demand position of the consumer market. All fishermen generally send the product to most popular market like Howrah Fish Market (Kolkata) and Shivaji fish market (Mumbai) etc. and as there is glut, price become down. Due to lack of information, fishermen are not sending their products in other fish market where they can get more prices. Fishermen also do not know that customers are demanding for when fish/fish product, so they are unable to satisfy the demand.

## Malpractices

Malpractices such as fraudulent weights, rigging of price etc and due to these, fish markets loose their good reputation and buyer think twice to go for the product. As there is no fixed price of fish, sometimes fishermen charge much higher price to new consumers, and loose their reputation.

## Less Popularity and Lack of Advertisement

Fish and fish products are not popular throughout India. Very less percentage of people are having fish as their daily food item. Only the people of coastal belt basically taking fish as food. This hampered the expansion of market.

Advertisement and popularization of fish products and their cost is negligible. People are not aware of fish, fish product, their grade, price, availability and importance.

## Problem of Middlemen and Lack of Cooperative Marketing

Cooperative marketing provide credit, market information, safety, fishing equipment, price distribution, helping in transportation of fish, boat repairing etc. But in Indian fish market the cooperative marketing is negligible.

*Chapter 20*

# Socio-economic Status of Fishermen

Several studies exist on the economic and social status of marine fishermen. However, such studies are limited in the case of reservoir fishermen. The problem for the development of reservoir fishermen is often looked into from a narrow angle all over the country. Hence, this chapter is an attempt to bring to light the socio-economic status of reservoir fishermen.

The studies of Raghavachari (1983), Devaraj and Mahadeva (1990), Devi (1997) George (2001) and Sakhare (2003) highlighted the socio-economic status of fishermen around different reservoirs in India.

Sakhare (2003) documented the socio-economic status of fishermen around Yeldari reservoir in Maharashtra.The age of active fishermen ranges between 18 to 56 years. In order to understand the position of fishermen in respect of education and literacy, separate information was obtained for males and females belonging to different age groups. Among these broad groups, an individual might either be an'educant' or a 'non-educant'. During the present investigation, the individuals were characterized as 'educants' who were receiving education at the time of survey. Among the noneducants were included all those who either had received education up to a certain standard, or were just literate or had no education at all. Of the total number of 189 males, there were as many 87 illiterates, 5 mere literates and a total of 29 who had received some education. Of the remaining 38 were school going *i.e.*, educants, and belonged mainly to age groups 5-24, and 30 were non-school going. The proportions for these five categories thus come to 46.03, 2.64, 15.34, 20.10 and 15.87 respectively. This shows the extremely poor progress of education.As regards females, the position on the basis of above categories was even worse. Of a total number of 121 females, there were as many as 71 illiterates (of which 24 were of non-school age), only 9 literates, 18 with some education to their credit and 23 school going. These give proportions of 58.67, 19.83, 7.43 and 14.87 respectively. The preponderant majority of females as well as males were thus illiterate.

The fish catch per day was calculated with the help of data obtained through questionnaire prepared in vernacular language. In Yeldari reservoir the total catch per fisherman varies from 0.4 to 3.0 kg. The average fish catch per day per fisherman was calculated at 1.56 kg only.The maximum catch was observed during monsoon, followed by winter and summer season. Majority (50 per cent) had a per capita monthly income range of only Rs. 1000 to 1100, 30 per cent of the fishermen come under income group of 500 to 600 followed by 20 per cent in range of Rs. 1200 to 1300 month.

The houses are small size, hut like. The walls are made-up of mud. The walls are plastered with dung and in some cases the wall is painted with white lime. Smoke outlets are not found in all the houses. Same room is used for sleeping, cooking and keeping domestic animals like cows, buffalows, and goats. Only in about 5 per cent of the houses, the domestic animals are kept in separate rooms. There is hardly any furniture in the houses. Traditional mud stoves called 'chul' are used for cooking. Build in mud stoves, surrounded on three sides by mud in an elevated platform are seen in all the houses. There is no separate kitchen room in most of the houses. The only source of water for domestic purpose is reservoir water, which is not treated before use. There is no water treatment plant or the water filter. The bathing and washing is also done on the bank of reservoir. Firewood collected from the forest is the only source of fuel. Twinges and branches collected from the forest is the only source of fuel. Sometimes trees are also fell for firewood. All houses depend on firewood collected from the forest and dung cake called 'Gouri' is also used as fuel. Agricultural activities are carried out throughout the year. Various crops are being raised according to the season. The crops cultivated are of two types *i.e.*, kharif and rabi crops. The former comprises of yellow and kharif jawar, pulses, cotton and groundnut. The important rabi crops are jawar, wheat and gram. Among the annuals, sugarcane, bananas and chillies are the most important. Majority of the cultivated area is under food crops and very less is under cash crops. Cultivated lands are irrigated by the electric motors which pumps the water from reservoir. Most of the lands are rain fed. Cattle dung used as the organic manure. Chemical fertilizers and pesticides are not in practice due to their high cost. Most of the fishermen have bulls for ploughing their lands. Cows and buffallows are very few in number and are the chief source of milk. Some of the householders rear goats and poultry for milking, meat and egg, the economy of all the fishermen solely depends on fishing and agriculture. The people owing the land cultivate various crops depending on the season and the products are utilized mostly for their own domestic use. The fieldwork is done almost by all the family members. The landless fishermen work as agricultural labourers.

Few of them work for carpentry. It is mainly a servicing industry, which carries out repairs of the wooden implements of the farmers, in return for a payment known as 'baluta' (It is a traditional form of payment in kind made by the cultivators to the village artisans and servants in return for the services rendered to him.) The raw material was supplied by the farmer and the wages of the carpenter were paid in kind based on the time-honored practices. If any new implements were made, the farmer had to pay for them. The payment, which was generally in kind varied from village to village for different implements.

Fishery potential of reservoirs is largely under exploited. Social problems arising from the exploitation of fishermen by middlemen act as impediments in improving their lot. The cooperative societies have a great role to play in the reservoir fisheries in India. Besides, leasing systems, machinery, stock monitoring and fishing efforts backed by efficient post-harvest management etc; can go a long way in improving the socio-economic status of small-scale fishermen operating in the reservoirs.

In Malampuzha reservoir of Kerala about 79.17 per cent of respondents were literate, though the literacy is largely confined to primary level (45.83 per cent). Both agriculture and fisheries provide work during the season and most people, both men and women, shift from one type of job to another based on season (Braj Mohan, 2001).

The total catch per fisherman of Yeldari reservoir varies from 0.4 to 3.0 kg. The average fish catch per day per fisherman was calculated at 1.56 only. The majority (50 per cent) of the fishermen had a per capita monthly income range of only Rs. 1000 to 1100, 30 per cent of the fishermen come under income group of 500 to 600 month, followed by 20 per cent in range of Rs. 1200 to 1300 month. The houses are small and hut like with walls made-up of mud. The economy of all the fishermen solely depends upon on fishing and agriculture.

Devaraj and Mahadeva (1990) accounted the socio-economic status of fishermen around Hemavathy reservoir in Karnataka. Fishermen usually engaged in fishing activities in groups during the peak season from May to January. During off-season some of them work as agricultural labourers and a very few of them work in their own agricultural land. The study revealed the following status: persons per family 2 to 5, male population among adults 57.14 per cent, active fishermen 47.23 per cent, the youngsters; women and children 42.15 per cent, and persons above 51 years of age 40.62 per cent. Majority of them were illiterate and had no exposure to modern fishery practices. Literacy rate was as low as 15.7 per cent. Not more than 10 per cent of children attended schools. The percentage of earning members was 60.81. During off season most of the male and female members work as agricultural labourers. Majority (76.85 per cent) had a per capita income range of only Rs. 500 to 1000 and came under the income group of Rs. 1000 to 5000 per family. A portion (19.22 per cent) of families belonged to the group that earned from Rs. 5000 to 9000 per annum with a per capita income of Rs. 1000 to 1500. Those who had income from both agriculture and fishing (3.93 per cent) earned more than Rs. 10,000 per annum with a per capita income higher than Rs. 1500. Fishermen were found to be engaged in house hold works and mending and making fishing nets. The main reason for the socio–economic backwardness of fishermen communities are the distribution of the communities in small clusters, inadequate number of nets and lack of proper understanding of the fishery.

According to Raghavachari (1983) the fishermen currently get about 10 to 15 per cent of consumer rupee. Very often they have to depend on contractors or commission agents for their fishery requisites. In the area of marketing infrastructure, basic facilities need to be provided. There is also an inadequacy of transport infrastructure for local and outstation sales. In addition inadequacy of ice plants and cold storages poses a problem for preserving fish. The government ice plants and cold storages are not utilized adequately or are not in working condition. There is also inadequate locational analysis while building markets and establishing ice plants and cold storages. In the area of marketing, the problems are the long channel of intermediaries leading to lower share of fishermen and higher price for the final consumer. The alternative systems of marketing could be through cooperatives and corporations. Past experience has shown that cooperatives have not been well organized, nor properly guided, and not efficiently lead so as to undertake marketing operations in the fisheries sector. The fishermen cooperatives societies do not seem to be active. These are more production based rather than marketing oriented. They have to depend on the department or corporation or contractor to market their catch.

Devi (1997) reported the socio-economic status of fishermen engaged in fishing of two minor reservoirs of Hyderabad. The fisherwomen populations are about 1000 with 200 households. The survey indicated a total of 843 with 21. 95 per cent males, 22.54 per cent; females and 55.52 per cent children. Percentage of male child is 64.53 per cent, female child is 35.47 per cent. The percentage of literacy was found to be considerably high at 74 per cent. The survey showed that almost all the respondents owned houses, either pucca, or semi pucca house or kutcha houses. The sources of drinking water are quite good as they have tap connections.

According to George (2001) if it is assumed that the fishermen depend only on fishing for their livelihood, it can be seen that except in Gobindsagar, in all other reservoirs of Madhya Pradesh they come under the category of destitute whose income is below Rs. 16000 annum. The fishermen of Gobindsagar reservoir can be grouped in the lower middle class whose income range in between Rs. 15000 to

45000. Infact reservoir fishery in all the cases mentioned above is only a supplementary means of livehood to the fishermen who also work as agricultural laborers and construction workers during period of low production and closed season.

Velayudhan and Sidharthan (1998) estimated the average annual gross income per full time fisherman from Chulliar reservoir was Rs. 12781. Due to increased fish production more quantity of fresh fish was made 'available to the rural consumers to meet their nutritional requirements. Surplus fish was transported to the distant urban markets which provided additional employment opportunities to rural people by way of handling, packaging, transport and marketing.

According to Sinha (1998) earnings of fisher folk is lower than unskilled farm workers though risks and uncertainties in catch and incomes are much more in their case. It should be better, if a cluster of reservoirs is taken for development so as to have a co-coordinated approach with regard to pre-harvest and post-harvest management policies. On trial basis management of few reservoirs may be handed over to fishermen's cooperatives under the supervision and guidance of state fishery officers. The result of the exercise will set the trend for future course of action.

## REFERENCES

Braj Mohan, 2001, Economic status of fishermen of Malampuzha feservoir. Paper presented during the *Nat. Sem. on Riverine and Reservoir Fisheries: Challenges and Strategies*, May 23 to 24, Cochin.

Devaraj, K.V. and H.S. Mahadeva, 1990. Socio-economic status of fishermen around Hemavathy reservoir, Karnataka. p. 102–105. In: Jhingran, Arun G. and V.K. Unnithan (Eds). *Reservoir Fisheries in India. Proc. of the Nat. Workshop on Reservoir Fisheries*, 3–4 January, Spl. Publ. 3, Asian Fisheries Society, Indian Branch, Mangalore, India.

Devi, Sarla, B., 1997. Present status, potentialities, management and economics of fisheries of two minor reservoirs of Hyderabad. *Ph.D. Thesis*, Osmania University, Hyderabad.

George, Ninan, 2001. Fishery management practices of the major reservoirs in Madhya Pradesh. Paper presented during *Nat.*

*Sem. on Riverine and Reservoir Fisheries: Challenges and Strategies*, May 23 to 24, Cochin.

Raghavachari, M., 1983. Reservoir Fishes in India: Some Issues and Problems. In: U.K. Srivastava and M. Dharma Reddy (Eds). *Fisheries Management in India: Some Aspects of Policy Management*. Concept Publishing Company, New Delhi, pp. 459–461.

Sakhare, V.B., 2003. Socio-economic status of fishermen around Yeldari reservoir, Maharashtra. *Aqua Tech.*, 2(5): 77–78.

Sinha, M., 1998. Policy options for integrated development of reservoir fisheries: from production to marketing. *Fishing Chimes*, 18(1): 54–59.

Velayudhan, T.D. and T. Sidharthan, 1998. Influence of stocking on fish yield in Chulliar reservoir and its impact on rural economy. *Fishing Chimes*, 18(4): 17–20.

*Chapter 21*

# Crafts and Gears Used in Fisheries

## Dugout Canoe

These are smaller types of boats and are used in the estuary, backwaters and rivers. These are cut out of a single log. These are known by different names like Danga, Dhoni, Vanchi, Hodi, Thoni and Ekhta. The size varies from about 3.5m to 14m in length. Generally the mango timber constructs it.

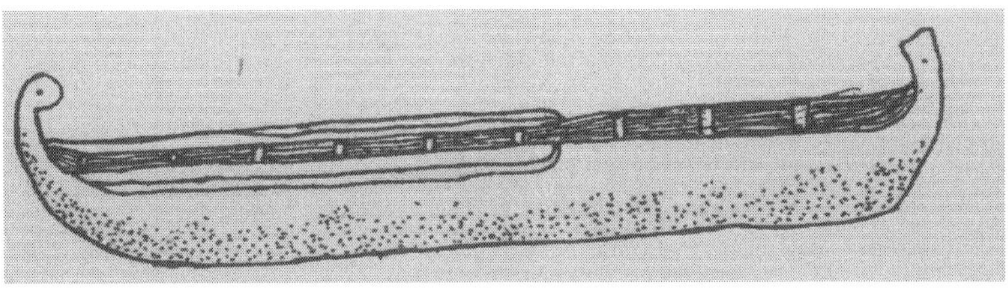

**Figure 21.1: Odam-A Dugout Canoe of Malbar**

## Raft

It is a simple craft. These are also known as 'catamarans'. The name 'catamaran' is originated from the Tamil word 'kattu maram' (kattu=binding, maram=woods).It is a primitive craft, constructed by tying together several logs which are curved and shaped like a cone. One end of the craft is shaped into a cone, which rises above the water level and from the point from where rudder is controlled. The size of catamaran varies according to the number and size of logs used for its construction. The size of catamaran varies from 9-12 m in length, width 0.7to 1.4m and depth 0.3 to 0.7m.

## Machwa

The size of this craft is about 9-14 meters in length. These are mostly used along Karwar on central west coast of India. It is mainly used for the transportation of cargo. In recent years these crafts are restricted themselves to some parts of west coast of India. It is a plank built craft. In recent years these have installed by inboard engines to enhance their speed and being used for fishing purposes also.

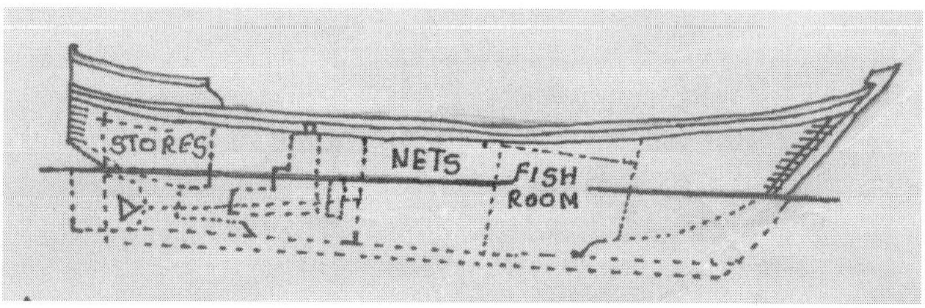


**Figure 21.2: Fishing machwa of Saurashtra**

## Masula

It is extensively used on the Coromandel Coast. It is about 9-12 meters or less in length. It is constructed with planks sown together with coir ropes and usually without frames or ribs. There are several variations of this type. In Orissa it is known as 'bar boats' and in Andhra Pradesh 'Padava' or 'Padagum'.

## Satpati

It is also known as 'Galbati'.It has a medium pointed bow, broad beam, straight keel and high gunwale. Satpati has been found to be an ideal type for mechanization as a motor engine can be fitted without any change in the design of the locally assembled boat.

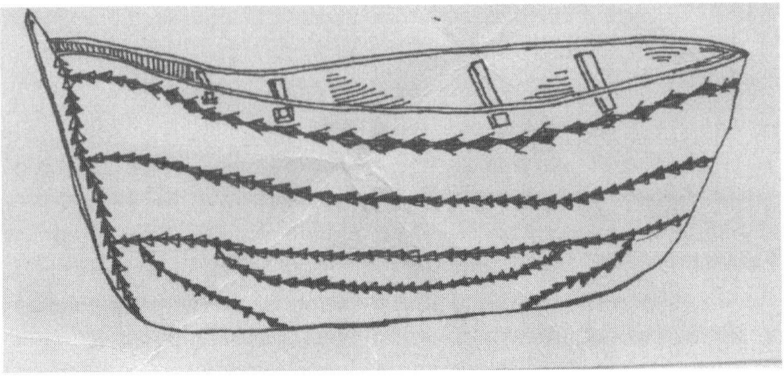

**Figure 21.3: Padava: A Masula Boat of Andhra Coast**

## Tuticorin

It is a carved model. Its common size is 11x2x1m.These are also known as 'fishing luggers' and are used more as mother ships and cargo boats than for purpose of fishing.

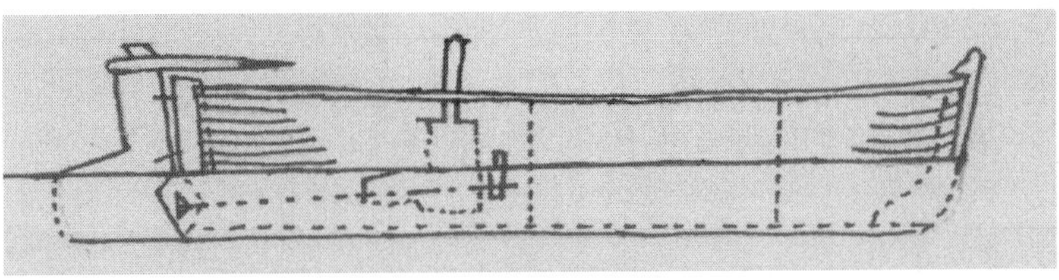

**Figure 21.4: Tuticorin Type Fishing Boat of Tamil Nadu**

## Gears Used in Fisheries

### Cast Net

It is a circular net of varying mesh sizes. It is known by variety of names. At the apex of the net a long thin rope is attached and at the periphery circular pockets are made by folding the net inwards. The pockets are provided with sinkers. The free end of the central rope is tied to the forefinger of the left hand. The net is then roughly folded in three equal portions, the fold near the rope is held by left hand, the middle one is kept free and the last one near the periphery,

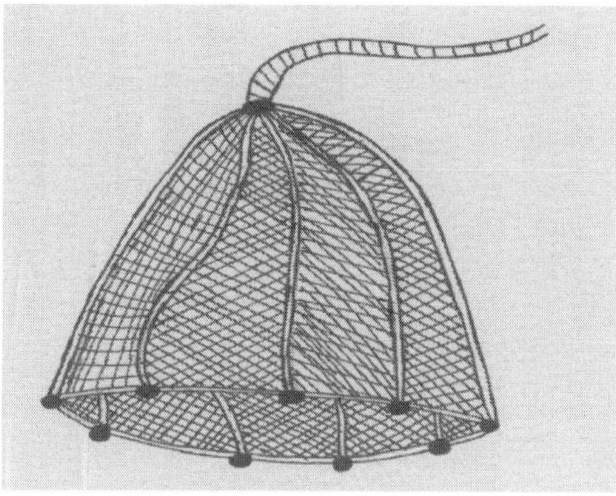

**Figure 21.5: Cast Net**

is lifted up over the head and thrown to a distance reachable by the attached rope. The net falls on the riverbed in a circle, trapping the fish in the area and ultimately into the pockets. The net is operated in shallow waters throughout the year especially for small fishes. Cast net with number of strings passively through the meshes from periphery to apex are also in use in some rivers.

## Gill Net

It consists of a single wall of webbing held vertical in water with floats and sinkers. Size of the mesh varies depending upon the size of the fish to be caught. A loosely hung net is more effective. When fish tries to swim through net, only its head passes through the mesh but its large body cannot. When fish tries to withdraw its head, twines slip under the gill cover and fish get gilled (caught) and hence the net is called as 'gill–net'.

Gill nets are set in two ways *i.e.*, stationary and floating. Stationary gill nets are set at the bottom with the help of stakes (poles) or anchors. These are used in shallow bays and estuaries. Floating gill nets are not firmly fixed to poles but are simply anchored on the bottom and are suspended with a float line and a lead line. Number of floats and sinkers are adjusted to set the net at the desired depth. This net is especially used in sea to catch herrings, Salmon, Mackerels etc.

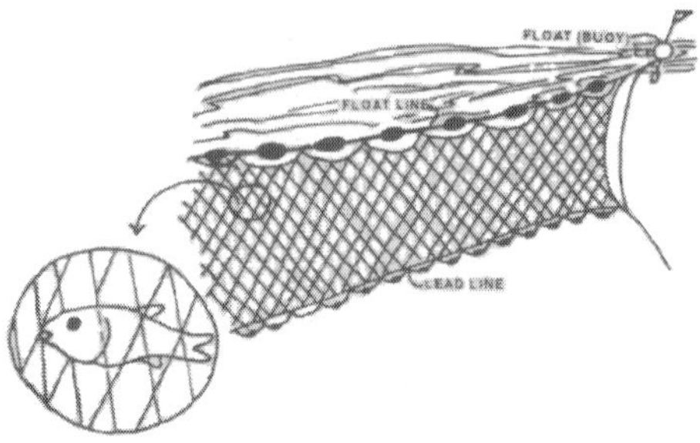


Figure 21.6: A Gill Net

## Drag Net

These are of large size and are of two types:

1. The first type consists of a bag with two wings and is known by several names as Pedda vala of telgu coast, Ber jal of Orissa and the Periya valaai of the Coromandel Coast.
2. The second type is a wall of enormous length. The upper margin of the net is supplied by a strong rope (head rope) and is provided with a large number of wooden floats. Along the lower margin is the footrope, to which are tied a number of weights or stone sinkers, to keep the net in position.

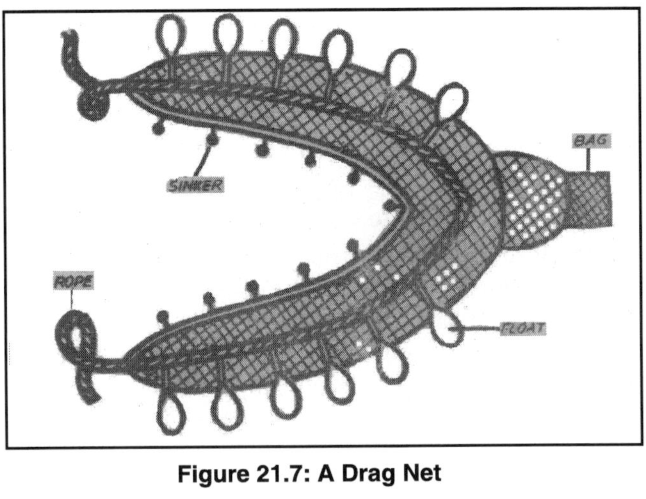


Figure 21.7: A Drag Net

Both types of nets are operated in a similar way. One end of the net remains on the bank of the river or sea shore, while a boat carries the rest of the net to spread it out in the water in a semicircular way, bringing the other extremity of the net, to another point on the river bank. The two ends of the net are then slowly dragged by two parties of fishermen.

## Dip Net

These are also known as lift net, and are of several shapes. These are lowered in to the water in the hope that fish would swim over them and lifted out of water. They are triangular, rectangular or square in shape and are fitted round a frame. Some bait is often put on the net or suspended over it to attract fishes. The smaller nets are operated by hand, while the larger ones are dipped and lifted from the water by means of a long pole, which is operated like a lever. The original simple hand dip nets have gradually been transformed into permanent stationary lift nets, operated from bridges made of bamboos. Depending on their design, the lift nets are of different types *i.e.*, Hela jal, Kharra jal, Bhesal jal and Khoursla jal.

## Hook and Line Gears

Hook and line gear is an ancient gear and is used to catch game or commercial fishes. The gear consists of a hook and a line. Metallic hooks of different shapes and sizes are used. Hook has several parts

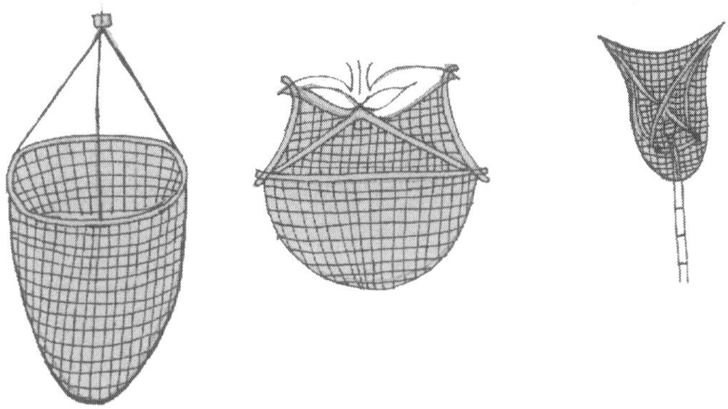

**Figure 21.8: Types of Dip Nets (lift nets)**

such as head or eye, shank, bend and spear. The spear end has an inclined barb with outer and inner points.

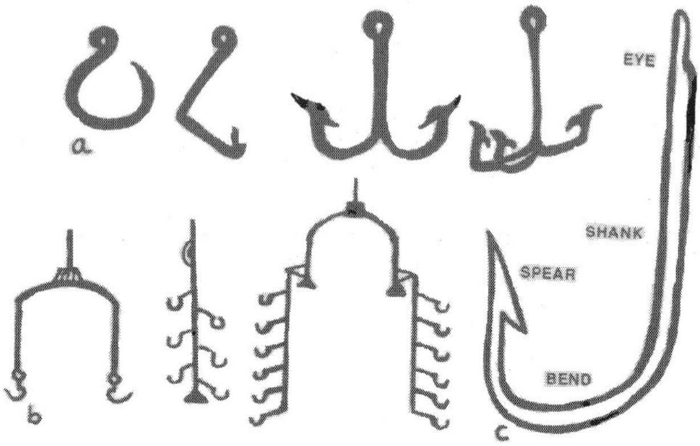


**Figure 21.9: (a) Various Forms of Hooks (b) Long Line with Several Hooks (c) Structure of Hook**

Hook and line fishing gears are of two types *i.e.*, hand line and long line. Hand line has a single vertical line with a sinker and a hook. It is used to catch a single fish. Long line consists of a long horizontal main line with several vertical lines bearing series of hooks. This is used to catch large number of fishes. For its operation, anchors, anchor lines and floats are used. The long line may be set stationary or towed through water or left drifting. For bottom feeding fish species, floating anchored type line is used. Before operating the gear, the hooks are baited. The fishes attempting to eat the bait get hooked and are then lifted. The bait is selected on the basis of type of fish to be caught. Generally prawns, earthworms, frogs, insects, small fishes and mixture of flour, baked rice and pieces of fish flesh are used as bait. Catch of fish depends upon many factors such as nature of bait, feeding behaviour of fish, size of hooks, abundance of natural food in water, density of fish and setting pattern of long lines.

# Index